STANDARD GRADE

BIOLOGY

Robert McMath

HODDER
GIBSON
AN HACHETTE UK COMPANY

Orders: please contact Bookpoint Ltd, 130 Milton Park, Abingdon, Oxon OX14 4SB. Telephone: (44) 01235 827720. Fax: (44) 01235 400454. Lines are open 9.00–5.00, Monday to Saturday, with a 24-hour message answering service. Visit our website at www.hoddereducation.co.uk. Hodder Gibson can be contacted direct on: Tel: 0141 848 1609; Fax: 0141 889 6315; email: hoddergibson@hodder.co.uk

© **Robert McMath 2005, 2009**
First published in 2005 by
Hodder Gibson, an imprint of Hodder Education,
An Hachette UK Company
2a Christie Street
Paisley PA1 1NB

This colour edition published 2009

Impression number   5 4
Year   2012

Cover photo © Art Wolfe/Science Photo Library
Artworks by Jeff Edwards
Cartoons © Moira Munro 2005, 2008
Typeset in 10.5 on 14pt Frutiger Light by Phoenix Photosetting, Chatham, Kent
Printed in Dubai

A catalogue record for this title is available from the British Library

ISBN-13: 978 0340 973 943

CONTENTS

Introduction 01

Chapter 1   **The Biosphere**
Investigating an Ecosystem 03
How it Works 08
Control and Management 15

Chapter 2   **Investigating Cells**
Investigating Living cells 21
Investigating Diffusion 23
Investigating Cell Division 27
Investigating Enzymes 31
Investigating Aerobic Respiration 36

Chapter 3   **The World of Plants**
Introducing Plants 39
Growing Plants 41
Making Food 49

Chapter 4   **Animal Survival**
The Need for Food 57
Reproduction 65
Water and Waste 72
Responding to the Environment 77

Chapter 5   **The Body in Action**
Movement 80
The Need for Energy 84
Coordination 93
Changing Levels of Performance 99

Chapter 6   **Inheritance**
Variation 103
What is Inheritance? 106
Genetics and Society 114

Chapter 7   **Biotechnology**
Living Factories 118
Problems and Profit With Waste 121
Reprogramming Microbes 128

INTRODUCTION

This book has been written with one aim in mind: to help you pass Standard Grade Biology. The book is not intended as a textbook to be used in the classroom or laboratory, but as a revision aid to help you prepare for tests and examinations. Its aim is simply to set out in a clear and concise form what you have to know and understand in order to do well in Standard Grade Biology. Don't forget that in your examination you will also be expected to show certain problem solving skills, such as drawing or interpreting graphs and diagrams and drawing comparisons and conclusions. These skills will also be helped by a good knowledge and understanding of Standard Grade Biology, which can be gained by studying this book.

Be prepared – and good luck!

Advice to Users

How to use this book

Standard Grade Biology consists of seven topics:

1   The Biosphere
2   Investigating Cells
3   The World of Plants
4   Animal Survival
5   The Body in Action
6   Inheritance
7   Biotechnology.

In each chapter, General Level work is identified by an icon, e.g. G1, G2 etc. Related Credit Level work follows and is identified by C1, C2 etc.

Beside each G or C icon you will find a short piece of writing in **bold**. This is a very brief description of what you should be able to do after completing this piece of work. These are called **Learning Outcomes** and can be used as checkpoints for your final revision. As you go through the book, at each checkpoint ask yourself, 'can I do that?' Also check that you can answer the brief questions that you will find at the end of each sub-topic. If you can't, work through the section again. Remember, all Standard Grade Biology exam questions must be based on these learning outcomes. If you know and understand them, you know everything you have to know for the knowledge and understanding element of Standard Grade Biology.

How to revise

You probably think you know how to revise. Sorry, you probably don't! Here is a list of things **not** to do when revising.

Common Mistakes

1 Don't simply read through your notes or a textbook (or a revision book, like this one). After the first few minutes, your mind will wander and you will be wasting your time.

2 Don't spend hours revising without a break. This may impress your parents, but again, your mind will wander and it will not be time well spent.

3 Don't copy out jotter notes or pages from a textbook. This may impress **you**, but once more, it can be done almost without thinking and doesn't often help.

Hints and Tips

Here are things you **should** do.

1 Take a piece of scrap paper and as you work through a section of notes or pages from a textbook or revision book (hopefully this one!) scribble down key words, key phrases and doodle bits of diagrams. It doesn't matter if you can't read the scribbles later because you should throw them away. The very action of picking key words **focuses** your mind on the page in front of you just as a lens focuses light.

2 Work in short sharp sessions – no more than 45 minutes without a break. Quality revision of the type mentioned above is tiring, but **quality** is much more important than **quantity**.

3 Set yourself targets, both short term and long term. Plan how you are going to revise all of your subjects and for each one, plan your revision strategy. For example, decide how many hours you are going to spend on The Biosphere, how many on Cells etc.

Each time you sit down to revise, set yourself a target of what you are going to cover. Plan a reward for yourself when you have reached your target – a bar of chocolate perhaps, or something less fattening!

4 Get hold of past Standard Grade exam papers and do as many as you can. But don't forget, they should be marked to make sure you are doing them right!

5 Don't panic!

Chapter 1

THE BIOSPHERE

Investigating an Ecosystem

Describing the ecosystem

G1 Identify habitat, animals and plants as the main parts of an ecosystem.

As well as studying the plants and animals (including humans), an investigation of an ecosystem must also include a study of the 'place' itself. For example, an investigation of a woodland ecosystem would have to include a study of the position of the wood. Is it sheltered or exposed? Is it on flat ground or on a slope?

The place where an organism lives is called its habitat. Within each ecosystem there may be a number of different habitats, each with its own wide variety of plants and animals.

Sampling organisms

G2 Give an example of a technique which might be used for sampling organisms and describe its use.

Sampling techniques are used to find out what (and approximately how many) animals and plants are present in an ecosystem, as it is usually impossible to count all members of a population.

Netting

Sweep nets with very fine mesh are swept through the branches of bushes and trees to catch small flying insects. These may be transferred to specimen jars for later identification.

Pond nets have coarser mesh to allow water to pass through easily while trapping small water organisms such as water beetles and water-dwelling insect larvae. In a stream, the net is held downstream and organisms, dislodged by kicking the bed of the stream, are carried into the net.

Quadrats

A quadrat is a square frame of a known area (usually 1 m^2). It is thrown at random in the area to be studied. The number of plants or animals of the type being investigated that are within the quadrat, are counted. This is repeated a number of times. The more times it is repeated, the more reliable the population estimate will be.

The population of daisies in a lawn could be estimated as follows:

Ten daisies were found in 6 m², therefore in 24 m² (four times as much) we would expect to find 4 × 10 = 40 daisies.

Count the daisies in the 'lawn' in Figure 1.1 to see how close the estimate comes to the true number. At school, you may learn alternative ways of using quadrats that give an abundance estimate, rather than an actual estimate of numbers.

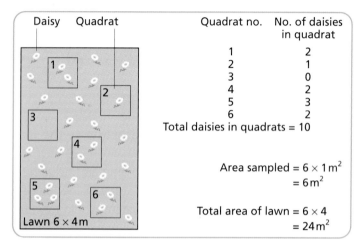

Quadrat no.	No. of daisies in quadrat
1	2
2	1
3	0
4	2
5	3
6	2

Total daisies in quadrats = 10

Area sampled = 6 × 1 m²
= 6 m²

Total area of lawn = 6 × 4
= 24 m²

Figure 1.1 **Sampling of daisies in a lawn**

Trapping

A pitfall trap can be used to catch crawling invertebrates from the leaf litter layer of a woodland floor or a field.

Small mammals can be live-trapped in baited traps made of metal or plastic.

Animals may be marked and released, unharmed, from these traps.

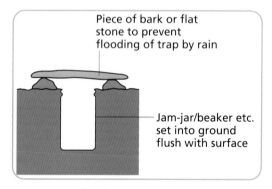

Piece of bark or flat stone to prevent flooding of trap by rain

Jam-jar/beaker etc. set into ground flush with surface

Figure 1.2 **Pitfall trap**

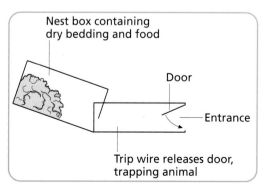

Nest box containing dry bedding and food

Door

Entrance

Trip wire releases door, trapping animal

Figure 1.3 **Mammal trap**

C1 Identify a possible source of error that might be involved in a sampling technique and explain how it might be reduced.

Errors in sampling, giving an estimate of a population which may be either too high or too low, can arise for a number of reasons.

1 Not enough samples taken. Go back to Figure 1.1 and work out the estimated population if only quadrats 1 and 3 had been counted. You should find that this gives an answer that is far too low.

2 Samples too closely grouped. This would not take account of any clumping of daisies in a lawn.

3 Deliberate placing of sample sites. Deliberately choosing 'good' sites would give too high an estimate of daisies on this lawn.

How to reduce errors in sampling

Errors can be reduced by:

1 taking an adequate number of samples.

2 avoiding the tendency to take samples from 'good looking' sites. This can be done by choosing sample sites at random, or by taking samples in a regular pattern.

Errors which may arise due to incorrect use of the sampling equipment must also be avoided. Care must be taken when using quadrats to count only those individual plants and animals which are at least half inside the quadrat. Care must also be taken to ensure that the organisms sampled are correctly identified.

Pitfall traps, mammal traps etc. must be set so that animals are able to enter the traps and must be inspected regularly.

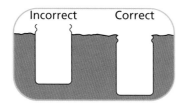

Figure 1.4 Pitfall trap

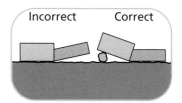

Figure 1.5 Mammal trap

Abiotic factors

G3 Identify two abiotic factors.

G4 Give an example of a technique which might be used to measure an abiotic factor and describe its use.

The distribution of plants and animals is affected not only by the other plants and animals (the biotic factors) present in their habitat, but also by non-living (abiotic) factors.

Table 1.1 Techniques for measuring abiotic factors

Habitat	Abiotic factor	Measurement technique
Woodland	Soil temperature	Soil thermometer/electronic thermometer
	Air temperature	Mercury thermometer
	Soil moisture	Soil moisture probe/weigh-dry-weigh soil sample
	Soil pH	pH probe and meter/pH indicator paper or universal indicator solution
	Light intensity	Light meter
Freshwater stream	Water temperature	Mercury/electronic thermometer
	Light intensity under water	Light meter with separate light probe
	Oxygen content of water	Oxygen electrode and meter/chemical analysis
	Flow rate	Float timed over measured distance

Recording errors

C

C2 **Identify a possible source of error that might occur during the measurement of an abiotic factor and explain how it might be reduced.**

Common errors of this type are:

1 failing to give thermometers long enough for the mercury to fully rise/fall

2 failing to allow meter needles to stop moving before taking a reading

3 casting a shadow over a light meter.

Errors in making comparisons of a number of sites

1 Weather conditions or time of day may have been different when sites were investigated.

2 Probes (e.g. soil thermometer, soil moisture etc.) may not have been pushed into the soil to the same depth at different sites.

How to reduce errors in recording abiotic factors

Avoiding these errors is done simply by correctly carrying out the techniques used.

Example

1   Allow thermometers and meters (pH, moisture, light, etc.) to stabilise before taking a reading. (Make sure you use the correct scale.)

2   Avoid casting a shadow over a light meter.

3   When making a comparison of different sites, make sure readings are taken under similar conditions, e.g. time of day, similar weather etc.

4   Make sure probes (temperature, moisture etc.) are handled in the same way at each site.

5   Take a number of readings at each site and work out an average.

The distribution of living organisms

G5 State the effect an abiotic factor has on the distribution of organisms.

The distribution of plants and animals is caused by many different factors. Climatic factors ('weather') have a major influence on plant and animal distribution. These non-living, or **abiotic** factors, include:

temperature

rainfall (or soil moisture)

light intensity

wind (direction and strength).

Although Scotland is a small country, different parts of Scotland will experience each of these abiotic factors to a different extent. For example, the south-east tends to be warmer and sunnier than the north-west. The west coast has higher rainfall than the east coast. The Outer Hebrides are more exposed to wind than the mainland.

On an even more local scale, each of these abiotic factors will vary over distances of only a few metres. A woodland floor will be cooler and darker than a cornfield only 20 metres away. A hilltop will be windier than a valley, and the valley soil will probably be deeper and damper than the hilltop soil.

All of these factors will determine which plants are able to grow in a particular place and the plants, in turn, will help to determine which animals can live there.

The effect of abiotic factors on distribution of organisms

C **C3 Explain ways in which abiotic factors can influence the distribution of organisms.**

Without detailed study of an organism and its habitat, it is only possible to suggest ways in which abiotic factors might effect its distribution. This is because a number of factors will be involved, not just a single one.

Example

1   The simple green plant *Pleurococcus* grows on the wettest side of a tree trunk but wind direction (blowing against the tree) and light intensity (without which the plant cannot photosynthesise) will also influence its growth.

2   The common dogwhelk, found on rocky shores, is only found where winter sea temperatures are between −1°C and 19°C. On each shore where they are found, however, their local distribution will be affected by wave action (abiotic factor) and availability of food (biotic factor).

Questions

G 1   The following list of environmental factors may affect trees growing on a hillside. Arrange them into (a) biotic factors and (b) abiotic factors.

   wind speed; number of leaf-eating insects; rainfall;

   temperature; number of deer; soil pH

C 2   List four abiotic factors that could affect distribution of single-celled green algae growing on a north-facing tree trunk.

 ### How it Works

G **G1 Describe what is meant by the words habitat, population, community and ecosystem.**

The word **habitat** was introduced at the beginning of this chapter. It means the place where a plant or animal lives. Some other important terms used in this chapter are:

Population: All the members of the same species living in a particular area.

Community: All the populations of plants and animals in a particular area.

Ecosystem: The community, plus the non-living environment in a particular area.

G2 Describe what is meant by the words producer and consumer.

The plant populations in a community can use the energy of sunlight to make their own food. Because they do this, plants are called **producers**. Animals cannot make their own food and so must eat plants or other animals. They are therefore called **consumers**.

Summary: Green plants are **producers**.

Animals (and non-green plants such as fungi) are **consumers**.

Herbivore = a consumer which eats only plants.

Carnivore = a consumer which eats only animals.

Omnivore = a consumer which eats both plants and animals.

G3 Give an example of a food chain and a food web.

G4 State that the arrows in a food web diagram show the direction of the flow of energy.

Food chains

A food chain is a simple way to describe how members of a community depend upon one another for food. For example, on the grasslands of the African savannah, zebras and lions, along with the grass, form a simple community.

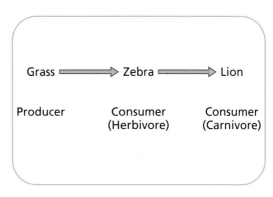

Figure 1.6 **Simple food chain**

The arrows in a food chain always point from the organism that is being eaten, to the eater. This is because the arrows show the direction of the flow of energy through the chain.

Although it is not usually shown in a food chain, the sun provides the energy for the plant.

A food chain must always start with a green plant.

Food webs

Food chains are very simple and do not show how complex the feeding relationships in a community really are. For example, in the food chain in Figure 1.6, zebras do not only eat grass; there will be other green plants in their diet. Lions do not only eat zebras; they will also eat gazelle, wildebeest etc.

A more complete picture of the feeding relationships in a community is given by a food web.

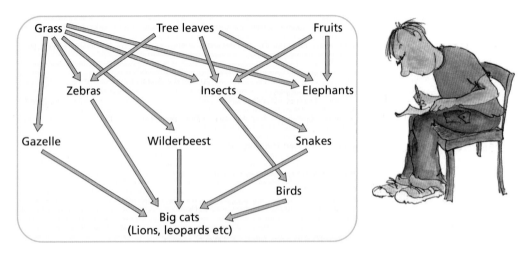

Figure 1.7 Part of a typical African savannah food web

Removing an organism from a food web

C1 Explain how removing one organism from a food web could affect the other organisms.

Removing all the thrushes from the following simple food chain would result in more snails surviving, which in turn would mean more grass being eaten.

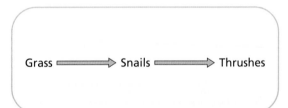

Figure 1.8 Simple food chain

Unexpected results may sometimes arise from the removal of an organism from a food web.

In Figure 1.9 of a food web, use of insecticide might remove the populations of insects and caterpillars. The carnivorous birds, having only snails now to eat, would decrease

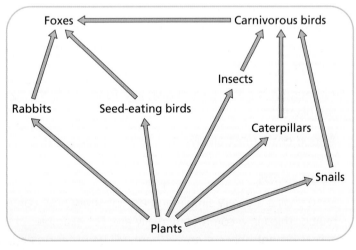

Figure 1.9 Food web

in numbers. Foxes would now have to eat more rabbits and seed-eating birds, so although rabbits and seed-eating birds might seem unlikely to be affected by insecticide, they may be, indirectly.

G5 State two ways in which energy can be lost from a food web.

When a plant or animal is eaten by another animal, only about 10% of the energy passed on is turned into new animal tissue. This huge energy loss at each stage in a food chain (about 90%) is due to food (chemical energy) being turned into or lost as:

1 heat energy

2 movement energy

3 energy lost in solid waste (the indigestible material).

Pyramid of numbers and pyramid of biomass

C2 Explain what is meant by the terms pyramid of numbers and pyramid of biomass.

A pyramid of numbers shows that the number of organisms at each level of a food chain decreases at each stage because of the large energy loss.

The pyramid of biomass is similar except that it represents the total mass of the organisms (not number) at each level in a food web. It is usually (although not always) the same shape as the pyramid of numbers.

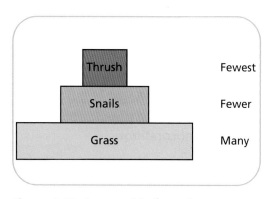

Figure 1.10 A pyramid of numbers

Population growth

G6 State that the growth rate of a population depends on birth rate and death rate.

If we assume that all members of a population stay in the area where they were born, then the population size depends on:

1 how many are being born (birth rate)

2 how many are dying (death rate).

If the birth rate is more than the death rate, the population will increase. If the birth rate is less than the death rate, the population will decrease.

What can limit the growth of a population?

G

G7 State three factors which can limit the growth of a population.

There are many ways in which growth of a population may be affected. The main factors limiting its growth are:

1 lack of food

2 lack of space

3 disease

4 predators.

If there is a lack of food, weaker individuals may die of starvation and fewer young may be produced.

Lack of living space may also reduce the number of young produced, as many animals need a certain amount of space for breeding. Overcrowding and lack of space may increase the level of disease in a population which may kill the weaker members.

As a population grows, there will be more food for its natural predators, therefore the population of predators will increase. Increased predation will decrease the population again.

Population graph

C

C3 Describe the shape of the growth curve of a population, under ideal conditions.

1 In a steady population the birth rate = the death rate. A graph would simply be a flat line.

2 For a population to increase (under ideal conditions) the birth rate must be much greater than the death rate.

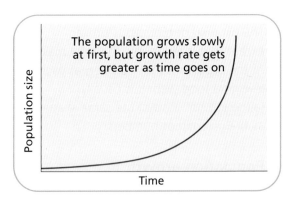

Figure 1.11 **Population graph**

C

C4 Explain the shape of the growth curve of a population, under ideal conditions.

This type of growth curve has three stages.

Growth is slow at first as reproduction is just beginning. Later, high birth rate and low death rate cause a rapid increase. Finally, several generations will be alive at the same time, all reproducing and the graph of population size rises almost vertically.

G8 State that competition occurs when organisms have a need for the same resources.

In all the examples given previously, animals (and often plants) are competing with others in the same population and in other populations for their basic needs: food, space, etc. (Examples in a plant population: sunlight, water, minerals from soil.)

Some examples of the effect of competition

G9 Describe some effects of competition.

Table 1.2 Effects of competition on organisms

Organism	Resource competed for	Effect of competition
Barley	Light	Plants grow tall to obtain maximum light
Barley	Soil minerals	Plants may be stunted or yellow if not enough minerals for all
Gannet	Breeding space	Those not obtaining a nest site will not mate
Gannet	Food	Youngest chicks may not survive if food is scarce

Nutrient cycles

G10 Explain why re-cycling of nutrients is important to the organisms in an ecosystem.

As well as energy, living things need minerals (e.g. nitrates and phosphates) in order to stay alive. Plants get these nutrients from the soil. Animals get them from eating plants or other animals (all originally coming from the soil).

Within an ecosystem, there is a limited supply of these nutrients. Plants are continually using them up and if they were not replaced, the soil would soon be unable to support any further plant growth. Fortunately, when plants and animals die, their remains are broken down by bacteria and fungi in the soil (**decomposers**)

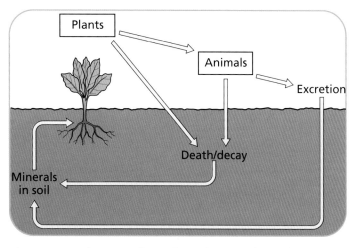

Figure 1.12 The re-cycling of nutrients in the soil

HOW TO PASS STANDARD GRADE BIOLOGY

and the mineral nutrients are put back into the soil once again. Also, animal excretion returns nutrients back to the soil.

Nitrogen cycle

C5 Describe the sequence of processes in the nitrogen cycle.

Nitrogen is an essential element for plants and animals, as it is necessary for the production of protein. It is re-cycled by the action of bacteria.

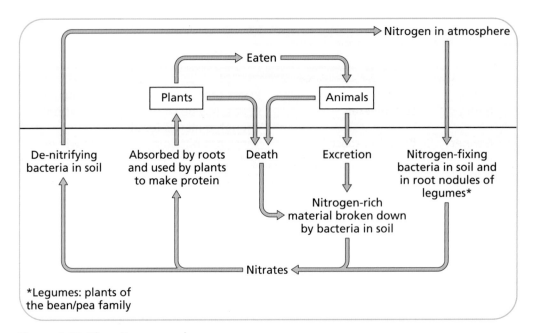

*Legumes: plants of the bean/pea family

Figure 1.13 **The nitrogen cycle**

Questions

1.   What do arrows in a food chain represent?

2.   Complete the following sentences:

   A food chain must always start with a _____ _____.

   The energy in a food chain is provided by the _____.

3.   Explain how each of the following factors may limit growth of a population: (a) lack of food, (b) lack of space.

Questions continued ➢

Questions *continued*

4 Explain the difference between a pyramid of numbers and a pyramid of biomass.

5 Why is nitrogen essential to plants?

6 Describe two ways in which soil nitrates may be produced.

Control and Management

Pollution

G1 State that pollution affects air, freshwater, sea and land.

Wherever we go, we find evidence of human activities polluting our environment.

The air is no longer pure; freshwater lochs, lakes and rivers are contaminated; the sea has become a dumping ground for our unwanted waste and the land contains man-made waste and impurities even thousands of miles from the nearest industrial areas.

The main sources of pollution

G2 State that the main sources of pollution are domestic, agricultural and industrial. Give an example of a pollutant from each category.

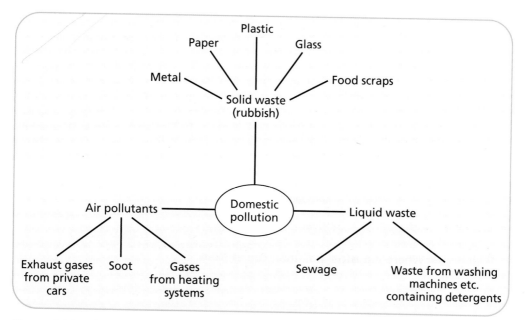

Figure 1.14 **Where domestic pollution comes from**

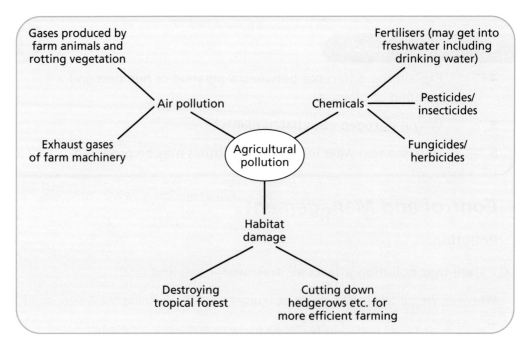

Figure 1.15 Where agricultural pollution comes from

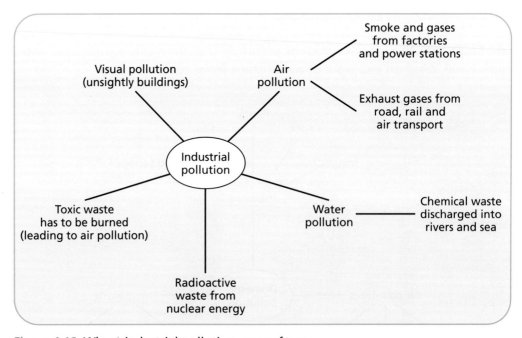

Figure 1.16 Where industrial pollution comes from

Fossil fuels

C1 Explain an undesirable effect of using (a) fossil fuels and (b) nuclear power as energy sources.

Fossil fuels (i.e. coal, oil and gas) all release undesirable chemicals into the environment when they burn.

These include:

1   sulphur dioxide which can lead to acid rain
2   oxides of nitrogen which can cause an increase in ozone which is desirable in the upper atmosphere, but damaging to crops at ground level
3   carbon monoxide which is poisonous
4   lead (car exhausts) which is known to cause brain damage.

Nuclear material, for example, waste released from nuclear power stations and taken into the food chain, may take thousands of years to disappear and may cause cancer in some of those exposed to it.

Pollution control

G3 Give an example of one way in which pollution may be controlled.

Some ways of reducing pollution are:

1   reducing the burning of fossil fuels by using alternative energy sources, e.g. wind power, solar power etc.
2   reducing the use of harmful substances and replacing them with 'environmentally friendly' ones, e.g. using aerosol propellants which do no damage to the ozone layer
3   re-cycling materials, e.g. glass, paper, metal. This saves raw materials and energy
4   cleaning up discharges to the environment, e.g. using filters to clean up car exhausts; 'scrubbers' to clean up discharges from factory chimneys.

Organic waste

G4 State that organic waste is a food source for micro-organisms.

Organic waste is composed of chemicals containing carbon or material of animal or plant origin, which has been released into the environment.

Most forms of organic waste can be used as food by micro-organisms such as bacteria and fungi.

Examples of organic waste are:

1 sewage

2 oil spills

3 sugar solution and other organic chemicals from industrial processes

4 blood from slaughter houses.

G **G5** Describe the effect of increased numbers of micro-organisms on the oxygen available to other organisms.

Most of the micro-organisms that use organic waste as food need oxygen as we do. In polluted water, growth of micro-organisms can lead to rapid use of the dissolved oxygen which is needed by the other water-dwelling organisms.

The effect of pollution on number of species

C **C2 Explain how organic waste pollution can affect the numbers of micro-organisms and hence oxygen concentration and numbers of species.**

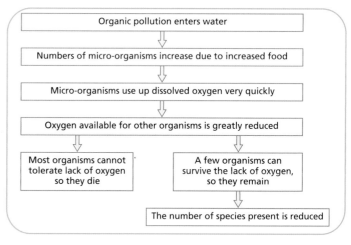

Figure 1.17 **Effect of pollution on the number of species**

Indicator organisms

C **C3 State what is meant by 'indicator species' and explain how they are affected by changing oxygen concentrations.**

Certain freshwater invertebrate animals are known to be very sensitive to changes in the concentration of oxygen in the water. If oxygen is used up by large numbers of micro-organisms feeding on pollution, they may not survive. So by being there or not being there, they can indicate how much pollution is in the water (for example, refer to Table 1.3).

Table 1.3 Effect of water pollution on organisms

	Organisms	Indication
A	Stone fly nymph May fly nymph	These animals only occur in unpolluted water
B	Caddis fly larvae Freshwater shrimp	If these are found without any group A animals, the water is slightly polluted
C	Water louse Blood worm	If these are found without any group A and B animals, the water has serious pollution
D	Tubifex worm Rat-tailed maggot	If these are found without any of the animals in groups A, B or C, it means the water is heavily polluted

Management of resources

G6 Give two examples of poor management of natural resources and suggest possible improvements.

G7 Describe how the effect of poor management of natural resources can lead to problems.

Table 1.4 Consequences of poor management of natural resources

Examples of poor management	Problems arising	Possible improvements
Destroying rain forest for short term gain from cash crops	Loss of species Possible climatic disturbance	Protect remaining rain forest Re-plant native hardwoods Provide financial incentives to stop growth of cash crops
Uncontrolled overfishing	Destruction of fish stocks (e.g. North Sea herring)	Regulate fishing effort Encourage catching of little-used species
Overgrazing of grassland	Soil erosion. Loss of fertile top soil	Rotate areas used for grazing
Over-use of chemicals for (a) pest control (b) improving soil fertility	Chemicals may harm other organisms in the food web Nutrients can pollute waterways	Reduce dependance on chemicals by encouraging more natural farming practices

Farm management and forestry

C4 Explain how parts of an ecosystem are controlled in either agriculture or forestry.

Farming

The parts of a farm ecosystem that a farmer must control are:

1   soil fertility (i.e. adequate essential minerals)
2   what plants are grown (i.e. only the desired plants – no competition from weeds)
3   disease-causing organisms.

Management methods

In the past, crop rotation was the only management method available. Crop rotation means growing different crops in successive years in a particular field, e.g. year 1 barley; year 2 sugar beet; year 3 clover (which replaces soil nitrates); year 4 barley again.

This system does not allow persistent weeds or disease-causing organisms to build up in the soil.

Nowadays, a combination of simple crop rotation plus agricultural chemicals is used.

Chemicals used:

1   artificial fertilisers (nitrates, phosphates and potassium)
2   insecticides to kill insects which may eat crops or carry disease
3   fungicides to prevent fungal diseases
4   herbicides to kill weed species.

Forestry

In addition to the factors of the ecosystem managed by farmers, foresters have a number of other factors to manage.

As forestry is often carried out on poor quality soil in exposed sites, seedlings are usually grown in protected nurseries, then transplanted out. Planting sites may need deep ploughing and drainage. Young trees need protection against animal damage, in particular by deer.

Due to the long growing period (70 to 80 years for conifers, 100+ years for hardwood) trees are planted close together then thinned at intervals, leaving the strongest specimens. Fire is another environmental hazard that must be managed by foresters.

Questions

1   Give an example of a pollutant from each of the following sources:
    (a) domestic; (b) agricultural; (c) industrial.

2   Give two examples of organic waste that can lead to pollution of freshwater.

3   Explain how organic pollution leads to a reduction in the number of species living in freshwater.

4   Explain what is meant by 'indicator species' and give a named example.

Chapter 2

INVESTIGATING CELLS

Investigating Living Cells

The 'bricks of life'

G1 State that cells are the basic units of living things.

With the aid of a microscope, it can be seen that all forms of life, from bacteria to the largest plants and animals, are all built up of common units ('building blocks') called cells. Cells are usually so small that they have to be magnified many times in order to be seen. The difference in size of living things depends largely on the **number** of cells in their body.

Unicellular organisms are organisms consisting of a single cell, e.g. bacteria, *Pleurococcus* (plant) and *Amoeba* (animal).

Very small objects such as bacteria are measured in μm, 1 μm = l/1000 mm

Simple multicellular animals (e.g. flatworms) and plants (e.g. moss), although very small, consist of hundreds of thousands of cells. Larger multicellular animals and plants consist of many millions of cells.

Generally, these cells range in size from 10 μm up to 200 μm.

Observing cells

G2 Explain the purpose of staining plant and animal cells.

Cells are viewed on a microscope by passing light through the specimen, then magnifying the image. The specimen must therefore be thin to allow light to pass through and, ideally, only a single layer of cells thick.

Some parts of cells are almost transparent and so they may appear almost invisible. In order to provide contrast, stain is added to the specimen. Some parts will absorb the stain better than other parts, so they will show up against unstained areas.

HOW TO PASS STANDARD GRADE BIOLOGY

Table 2.1 Two common stains

Material	Stain	Effect
Plant cells (e.g. onion epidermis)	Iodine	Stains nuclei brown
Animal cells (e.g. human cheek cells)	Methylene blue	Stains nuclei blue

Structure of cells

G3 Describe the structure of a typical plant and animal cell and list the differences between them.

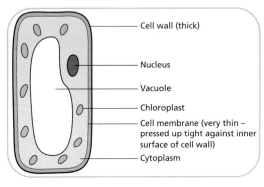

Cell wall (thick)

Nucleus

Vacuole

Chloroplast

Cell membrane (very thin – pressed up tight against inner surface of cell wall)

Cytoplasm

Figure 2.1 Typical plant cell

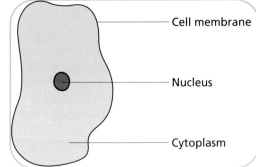

Cell membrane

Nucleus

Cytoplasm

Figure 2.2 Typical animal cell

Table 2.2 Differences between plant and animal cells

	Plant cells	Animal cells
Cell wall	Always	Never
Large central vacuole	Most	Never
Chloroplasts	Some (parts exposed to light only)	Never

As can be seen from Table 2.2, the surest way of distinguishing between a plant and animal cell is by the presence or absence of a cell wall.

The functions of the various parts of the cell are not dealt with here. We will cover this later on.

G

Questions ?

1 Name two stains that are used for examining cells and explain why their use is necessary.

2 Construct a table showing (a) similarities and (b) differences between plant and animal cells.

Investigating Diffusion

High to low

G

G1 State that a substance will diffuse from an area of high concentration to an area of low concentration (of that substance).

The sugar lump in the cup will slowly dissolve and the water molecules, moving in all directions, will bump into the sugar molecules and gradually spread them out.

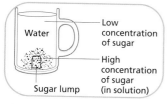

Figure 2.3 **Sugar lump in a cup**

After some time

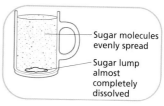

Figure 2.4 **Dissolved sugar lump in a cup**

The sugar molecules have **diffused** from an area of **high** sugar concentration to an area of **low** sugar concentration. Eventually, the concentration of sugar molecules will be the same everywhere in the cup.

Substances can also diffuse through a gas e.g. smelly gas escaping from a chemistry laboratory (high concentration) will spread by diffusion to other parts of the school (low concentration).

Diffusion and cells

G

G2 Give examples of substances which enter and leave the cell by diffusion, e.g. dissolved food, oxygen, carbon dioxide and water.

Oxygen and dissolved food are raw materials and will diffuse into an animal cell. Carbon dioxide is a waste product and will diffuse out of an animal cell.

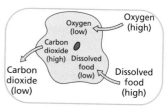

Figure 2.5 **Diffusion of substances in a cell**

i

Green plant cells in sunlight may behave differently – see 'The World of Plants', Chapter 3.

Diffusion in whole organisms

C

C1 Explain the importance of diffusion to organisms.

In the same way that gases move in and out of cells, oxygen and carbon dioxide also move in and out of whole organisms, both animals (including humans) and plants.

In small organisms (e.g. amoeba, bacteria etc.) gas diffuses in and out through the cell membrane.

In larger organisms, there are special organs for gas exchange. These are needed as diffusion would be too slow.

Table 2.3 **Gas diffusion in whole organisms**

Organism	Gas exchange 'organ'
Amoeba	Cell membrane
Tree	Leaves
Fish	Gills
Cow	Lungs

Diffusion through the cell membrane

G

G3 State that the cell membrane controls the passage of substances in and out of the cell.

Put very simply, a cell membrane has holes (or pores) of a certain size. Small molecules may pass through these pores, but large molecules may not.

The glucose and sucrose molecules are both at a higher concentration on the left of the membrane so

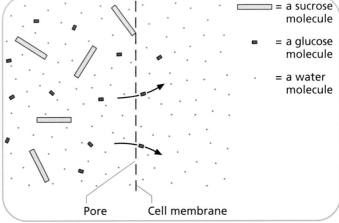

= a sucrose molecule

= a glucose molecule

= a water molecule

Pore   Cell membrane

Figure 2.6 **Diffusion through a cell membrane**

will try to move from left to right by diffusion. The glucose molecules are small enough to pass through the pores in the membrane, but the sucrose molecules are too large.

Osmosis – the diffusion of water through a membrane

G4 **Explain that osmosis is a 'special case' of the diffusion of water.**

Water molecules are very small and can pass through the pores in the cell membrane with no difficulty. This movement of water through a cell membrane from a higher water concentration to an area of lower water concentration is called **osmosis**.

Plant and animal cells (or even whole plants and whole animals) can gain or lose water by osmosis. The movement of water into or out of a cell depends on the concentration of the solution surrounding the cell.

Cells in a weak (very watery) solution

Plant cell in water or dilute solution (e.g. when a plant has adequate water available).

Water moves in by **osmosis**.

Cell swells up (becomes **turgid**) but cannot burst because of the tough cell wall. This is desirable in plants as it makes the whole plant more rigid.

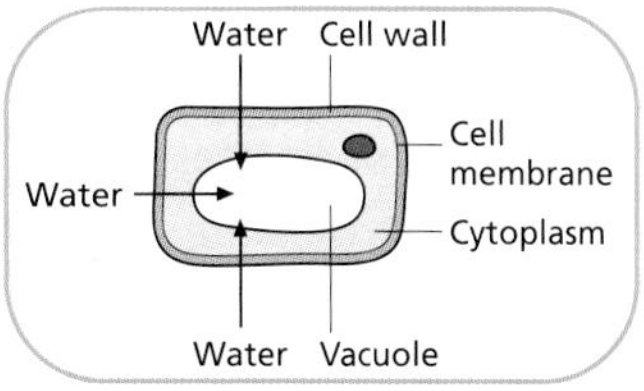

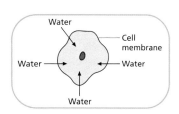

Figure 2.7 Plant cell in a weak solution

Animal cell in water or dilute solution (e.g. when blood becomes diluted due to drinking a lot of liquid).

Water moves in by **osmosis**.

Cell swells up and without a tough cell wall to protect it, it is in danger of bursting. This could be very serious in animals (e.g. red blood cells might burst) but excess water will be removed by the kidneys to stop this happening (see, 'Water and Waste').

Figure 2.8 Animal cell in a weak solution

Cells in a strong solution (less water)

Plant cell in strong solution (e.g. when plant has lost too much water).

Water moves out by **osmosis**.

Vacuole shrinks, membrane pulls away from wall and cell becomes flaccid. The plant will wilt. The cell is said to be **plasmolysed** and the process is **plasmolysis**.

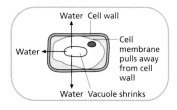

Figure 2.9 Plant cell in a strong solution

Animal cell in strong solution (e.g. when blood becomes too concentrated due to sweating).

Water moves out by **osmosis**.

Cell shrinks and it may not work properly. The animal suffers from dehydration.

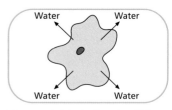

Figure 2.10 Animal cell in a strong solution

These examples show that water moves from one side of a membrane, where there is plenty of water, to the other side, where there is less water. Osmosis is therefore a special type of diffusion of water, through a membrane.

C C2 **Explain osmosis in terms of a selectively permeable membrane and of a concentration gradient.**

C C3 **Explain osmosis in terms of water concentration of the solutions involved.**

To help understand diffusion of water, try to imagine a high and a low water concentration, e.g. a 20% salt solution (A) must be 20% salt and 80% water and a 5% salt solution (B) must be 5% salt and 95% water. So we can think of solution A as having a lower water concentration and solution B as having a higher water concentration.

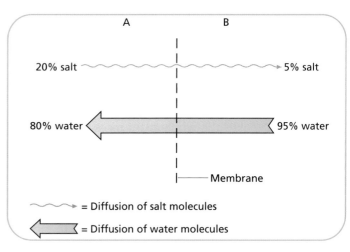

Figure 2.11 Movement of water and salt molecules

Now imagine these two solutions (A and B) separated by a membrane (see Figure 2.11).

Water molecules, being smaller than salt molecules can move through the membrane faster. Very large molecules cannot pass through the membrane at all. Because of this, membranes are referred to as **selectively permeable**. The difference in water concentration on the opposite sides of the membrane causes what is called a **concentration gradient**.

Remember, gradient = slope. Things will roll down a slope from high to low, so water moves down a water concentration gradient from a high water concentration to a low water concentration.

Questions

1 Copy and complete the following definition of diffusion:

A substance will diffuse from an area of _____ _____ to an area of _____ _____.

2 Construct a table showing which of the following substances enter and leave an animal cell by diffusion:

oxygen; carbon dioxide; dissolved food

3 What part of a plant cell controls entry and exit of materials?

4 (a) What is likely to happen to a red blood cell placed in pure water?

(b) Describe what happens to a plant cell placed in strong salt solution.

Investigating Cell Division

Why do cells divide?

G1 State that cell division is a means of increasing the number of cells in an organism.

New cells are needed by multi-cellular plants and animals for two purposes:

1 growth (for an increase in size)

2 replacement of dead cells (most cells live for a much shorter time than the whole organism of which they form a part).

There are special cells in both plants and animals that keep growing and dividing to produce new cells for these two purposes.

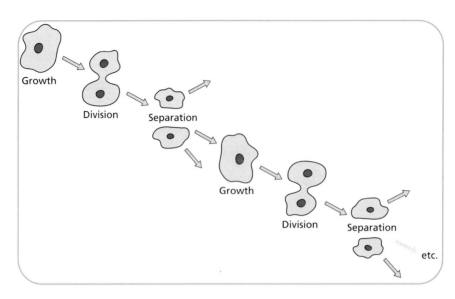

Figure 2.12 **Cell division**

The importance of the nucleus

(G) **G2 State that the nucleus of the cell controls cell activities, including cell division.**

The nucleus is the control centre of a cell. Inside the nucleus are a number of thin, thread-like objects called **chromosomes**. The number of chromosomes is the same in every cell of an organism's body. Each species has a characteristic chromosome number. The human chromosome number is 46 (i.e. there are 46 chromosomes in the nuclei of our cells).

The chromosomes carry a complete set of information relating to the particular individual in whose cells they are. Everything that a cell does, including cell division, is controlled by the information on the chromosomes. When a cell divides, it is essential that each of the two cells resulting has a complete set of instructions. That is why nuclear division must take place before a cell can split into two daughter cells.

Mitosis

G3 **State that each of the two cells produced by cell division has a complete set of chromosomes and the same information.**

If a human cell has 46 chromosomes, when it divides, the daughter cells must also have 46 each; but not just any 46. They must both have complete and identical sets.

Each one of the 46 chromosomes is copied and separated so that one of the copies goes into each daughter cell. This process of copying and separating the chromosomes, followed by division of the cytoplasm, is called **mitosis**.

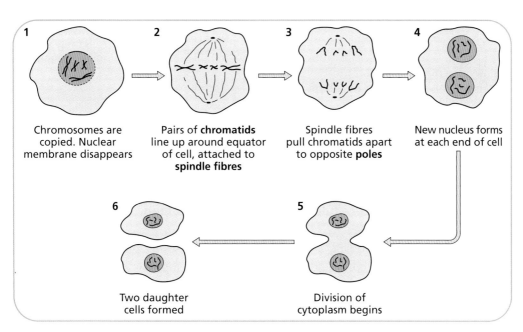

1	2	3	4
Chromosomes are copied. Nuclear membrane disappears	Pairs of **chromatids** line up around equator of cell, attached to **spindle fibres**	Spindle fibres pull chromatids apart to opposite **poles**	New nucleus forms at each end of cell

6	5
Two daughter cells formed	Division of cytoplasm begins

Figure 2.13 **Summary of mitosis**

Chromatids: the name given to the copied chromosomes, still joined together, before separation.

Spindle fibres: fibres of cytoplasm that pull the pairs of chromatids apart.

Poles: the opposite ends of the cell.

The main events in cell division

C

C1 **Describe the stages of mitosis.**

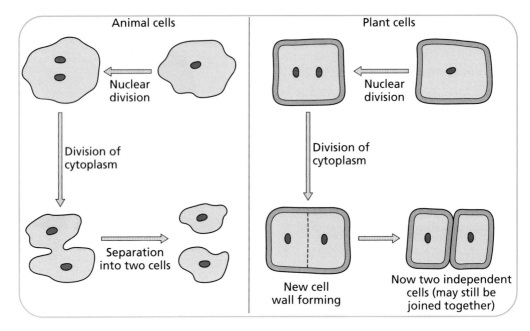

Figure 2.14 **The events in cell division**

Maintaining the chromosome complement

C

C2 **Explain why it is important that the chromosome complement of daughter cells in multi-cellular organisms is maintained.**

The reasons for the doubling and separating of chromsomes have been explained in G3. If the mechanism breaks down and a cell inherits too many, too few or an incorrect set of chromosomes, it may not behave as it should and the cells may die or grow abnormally. Problems in the division of sex cells can lead to even more serious abnormalities in the next generation, such as Down's syndrome (see 'Genetics and Society').

G

G4 Identify the correct sequence of stages of mitosis from a series of drawings or diagrams.

Using the information in Figure 2.13, sort out the following jumbled up diagrams of mitosis into the correct order.

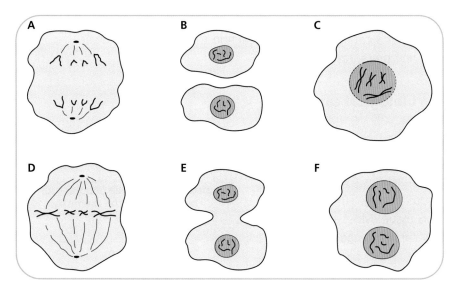

Figure 2.15 Jumbled up diagram of mitosis

Questions

1 A cell from a cat contains 38 chromosomes. Cells produced by mitosis in a cat will contain:

A 19 chromosomes;

B 38 chromosomes;

C 76 chromosomes.

2 Explain why it is important that chromosomes are not lost or gained during cell division.

Investigating Enzymes

What is a catalyst?

G1 Explain the meaning of the term 'catalyst'.

A catalyst is a chemical substance which helps a chemical reaction to take place, but is not used up in the reaction itself

e.g. chemical A + chemical B (+ catalyst) → chemical C.

Chemicals A and B might react under conditions of high temperature and pressure, but the addition of a catalyst causes the reaction to take place at lower temperatures and pressures.

Enzymes – biological catalysts

G **G2 Explain why enzymes are required for the functions of living cells.**

Thousands of reactions such as the one on the previous page (A+B→C) take place in living tissue. Most are controlled by catalysts made by the cells themselves. These biological catalysts are called **enzymes**.

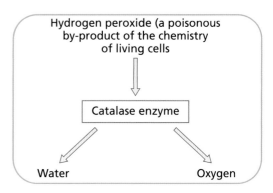

1   Hydrogen peroxide breaks down slowly by itself, but very much faster when assisted by the enzyme catalase.

2   The catalase is not used up in the reaction and can go on to break down further hydrogen peroxide molecules.

Figure 2.16 Example of enzyme action

Amylase, a 'breaking down' enzyme...

G **G3 Give an example of an enzyme involved in the chemical breakdown of a substance.**

The enzyme **amylase** is found in the digestive system of animals and in the seeds and food storage organs of plants. Amylase is responsible for the breakdown of large starch molecules into smaller sugar molecules.

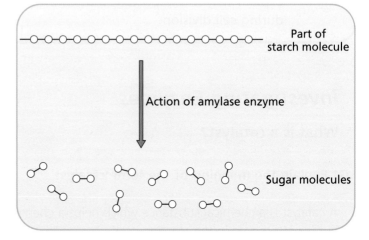

Figure 2.17 The action of amylase

...and potato phosphorylase, a 'joining together' enzyme

G4 Give an example of an enzyme involved in synthesis.

Many enzymes take small substrate molecules and join them together to make larger product molecules.

Substrate: the chemical substance on which an enzyme acts, either by breaking it down or building it up.

Product: the molecules that are produced as a result of enzyme action.

Table 2.4 **Enzyme action**

Substrate	Enzyme	Product
Starch	Amylase	Simple sugars
Glucose-1-phosphate	Phosphorylase	Starch

Sugar made in the leaves of a potato plant (see 'Making Food') has to be turned into starch for storage in the underground tubers (i.e. potatoes). This is done by the enzyme **potato phosphorylase** joining together molecules of the sugar **glucose-1-phosphate**.

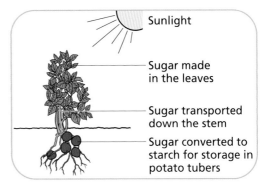

Figure 2.18 **Phosphorylase in action**

Enzymes are 'specific'

C1 Explain the term 'specific' as applied to enzymes and their substrates.

A key will only open a lock with a matching shape. In the same way, an enzyme molecule has a particular shape and will only affect substrate molecules with a matching shape. The enzyme is said to be **specific** to that substrate.

The example in Figure 2.19 is of a 'breaking down' enzyme in action, e.g. amylase. A 'joining together' enzyme, e.g. phosphorylase, works in reverse.

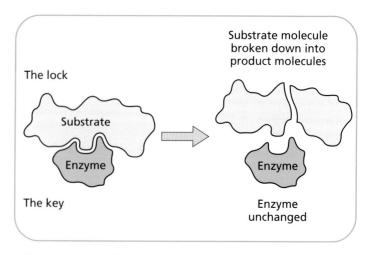

Figure 2.19 'Lock and key' model

The chemical nature of enzymes

G5 State that enzymes are proteins.

All enzymes are **protein** molecules. Acting as enzymes is one of the most important functions of proteins in living cells. The fact that enzymes are proteins explains their responses to temperature and pH covered in G6 and G7.

The effect of temperature and pH on enzyme activity

G6 Describe the effect of temperature on enzyme activity.

G7 Describe the effect of a range of pH on the activity of pepsin and catalase.

Enzymes are very sensitive to changes in cell temperature and pH.

Temperature
The effect of temperature on enzyme activity (i.e. how fast it is working) can be shown by a graph.

The graph shows that at low temperatures the enzyme works very slowly then speeds up as the temperature rises. In the case of human enzymes, most work fastest at about body temperature (37°C). As the temperature increases further, enzyme activity quickly decreases and stops altogether above about 50°C.

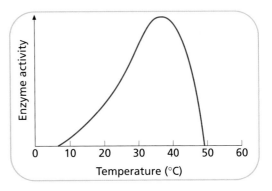

Figure 2.20 Temperature graph

pH

Every enzyme works best at a particular pH. The following graph shows how two human enzymes prefer different pH conditions.

The graph shows that pepsin (a protein-digesting enzyme secreted by the stomach) works best in acidic conditions. Catalase (see G2, p. 32) works best in alkaline conditions. The ability of pepsin to break down a protein such as albumen (egg white) is much reduced if the pH is not acidic enough.

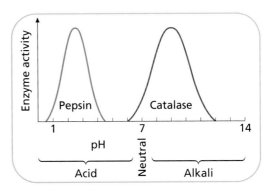

Figure 2.21 pH graph

Optimum conditions for enzyme action

C2 Explain the term 'optimum' as applied to the activity of enzymes.

The graphs in Figures 2.20 and 2.21, show that enzymes have a particular temperature and pH at which they work best. These 'best' conditions are called the optimum conditions for that particular enzyme.

Questions

1   Explain why enzymes are called 'biological catalysts'.

2   Copy and complete the following sentences:

(a) Amylase is an enzyme involved in the chemical _____ of _____ molecules into smaller _____ molecules.

Questions continued ➢

Questions *continued*

(b) _____ is an enzyme found in potatoes, which joins _____ molecules together to make larger _____ molecules. This is an example of a _____ (i.e. 'joining together') enzyme.

C

3 Explain why enzymes are said to be specific to their substrates. (Use the term 'lock and key' model in your answer.)

4 What is the optimum temperature for most human enzymes?

Investigating Aerobic Respiration

The need for energy

G

G1 State three reasons why living cells need energy and give an example of an energy transformation in a plant or animal.

Living cells need energy for:

1   cell division

2   cell growth

3   movement (muscle cells)

4   making certain chemical reactions work.

G

G2 State that cells need oxygen to release energy from food during aerobic respiration.

Burning uses up oxygen. The release of energy from food inside a cell also needs oxygen.

food + oxygen → energy + waste products

The release of energy from food takes place inside cells and is called **cellular respiration**.

i

As the process uses oxygen, it is referred to as **aerobic respiration. Aerobic =** with 'air' – as in aerobatics, but here, specifically with oxygen.

Summary

G3 Describe aerobic respiration in terms of a word equation.

food (e.g. glucose) + oxygen → energy + carbon dioxide + water

The waste products from aerobic respiration

G4 State that the carbon dioxide given off by cells during tissue respiration has come from food.

G5 State that heat energy may be released from cells during respiration.

Carbohydrates, proteins and fats, the three main types of food that are used by living cells, all contain the elements **carbon**, **hydrogen** and **oxygen**. As energy is released from these foods inside the cell, the carbon and oxygen are released in the form of **carbon dioxide** gas.

Some of the energy released from food during respiration is in the form of heat energy. It is possible to tell the difference between living cells and dead cells by measuring the heat given off by the living cells (e.g. germinating peas are warmer than dead peas).

The energy content of food

C1 State that fats and oils contain more chemical energy per gram than carbohydrates or proteins.

The energy contained in food can be estimated by burning. The heat given off by the burning food sample heats a known volume of water and the rise in temperature of the water is measured.

1 g of carbohydrate and 1 g of protein both give off approximately 20 kilojoules (units of energy).

1 g of fat (or oil) gives off approximately 40 kilojoules.

This shows that fats contain approximately twice as much energy as equal masses of carbohydrates or proteins.

Energy and cell metabolism

C C2 **Explain the importance of energy released from food during respiration, to the metabolism of cells.**

All the chemical reactions taking place inside a cell are known together as **cell metabolism**. Many of these reactions, without which the cell cannot survive, need energy to make them work:

e.g. A + B → nothing happens
but
A + B (+ energy) → C.

The energy to make these reactions take place comes from the chemical energy in food.

Questions

G

C

1   Give three reasons why living cells need energy.

2   Write the word equation that describes aerobic respiration.

3   Which of the following substances will contain most energy?

   A   5 g protein

   B   2 g sugar

   C   3 g fat

   D   4 g starch

THE WORLD OF PLANTS

Introducing Plants

Variety in the plant kingdom

G1 Give examples of advantages of there being a wide variety of plants.

G2 Describe three specialised uses of plants.

Humans make many different uses of plants. This variety of uses is possible because such a wide variety of plant types exist.

Here are a few examples:

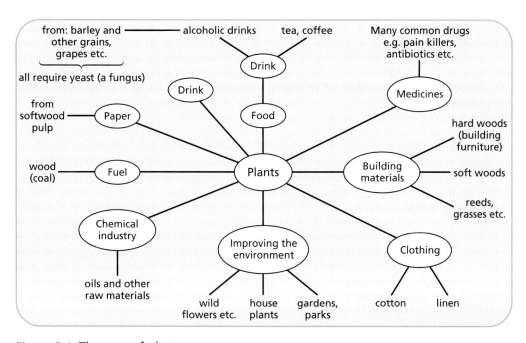

Figure 3.1 The uses of plants

Medicines, clothing and paper are three specialised uses of plants. They are specialised because the plants used as raw materials must be processed or treated in some way before the product is produced.

Extinction of plant species

 C1 Explain possible consequences to humans and other animals of a reduction in the variety of plant species.

Some plant species are being used up faster than they can be replaced. Certain hardwood trees used in the building and furniture industries are in danger of extinction.

Many other plant species are not used directly by humans, but their loss would mean a reduction in the 'storehouse' of plant material which may prove in the future to hold new food species, new medicines and new chemicals.

Yet other plant species may play a key part in the food webs of our planet. Their loss could have unexpected consequences for humans and other animal species.

Processing plants

C2 Describe a production or refining process e.g. malting barley, rape seed, raspberries, timber.

Many specialised uses of plants have been shown in Figure 3.1 on page 39. In almost all cases, the product is not the raw plant material (with the exception of some foods), but a processed or refined form of the plant material.

Malting barley

Barley is a raw material used in the manufacture of beer and whisky. Malting is the process of converting starch in the barley into sugar which is then converted into alcohol by yeast.

New uses of plants

C3 Describe two potential uses of plants or plant products (e.g. new medicines, new food sources).

New medicines

Many new medicines, particularly antibiotics, have been developed from chemicals obtained from plants.

New food sources

Some wild varieties of plants are known to be rich in nutrients such as protein and carbohydrate and may, in the future, be developed as commercial crop plants.

Questions

1   Give three examples of specialised uses of plants or plant materials.

2   Give an example of a possible new use of plants or plant materials.

Growing Plants

Structure of a seed

G1 Describe the functions of three main parts of a seed (i.e. seed coat, embryo and food store).

Most plants begin life as a seed. The structure of a seed protects and feeds the embryo plant inside.

Table 3.1

Part	Function
Seed coat	Protects the embryo plant and its food store until conditions are suitable for germination (growth)
Embryo	The future plant. Part will develop into roots and part into stem and leaves
Food store	Food (starch) to keep the embryo alive inside the seed and during germination until the new leaves can make more food

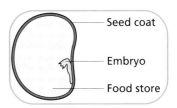

Figure 3.2 Inside a broad bean seed

HOW TO PASS STANDARD GRADE BIOLOGY

Conditions for germination

(G) **G2 Describe the effect of temperature and the availability of water and oxygen on germination.**

Germination is the process in which the embryo plant inside a seed begins to grow. Roots grow down into the soil and a shoot grows up through the soil to emerge into the light.

Seeds need three environmental conditions to be suitable before they can germinate.

1	Temperature	If temperature is too low, germination is prevented. (If it were not, frost could kill the emerging seedling.)
2	Water	Water is needed to soften the seed coat and start the growth of the embryo.
3	Oxygen	During germination, the growing embryo needs much more oxygen than it did while it was dormant inside the seed. Insufficient oxygen will prevent germination.

Temperature and germination

(C) **C1 Describe the changes in percentage germination that occur over a range of temperatures.**

If the temperature is too low, the enzymes involved in bringing about germination will work slowly or not at all. If the temperature is too high, the enzymes may be destroyed and germination will be prevented. The highest percentage of seed germination will occur at the optimum temperature for the enzymes involved.

Flower structure

G3 Describe the functions of the parts of flowers (i.e. sepal, petal, stamen, anther, stigma, ovary and nectary).

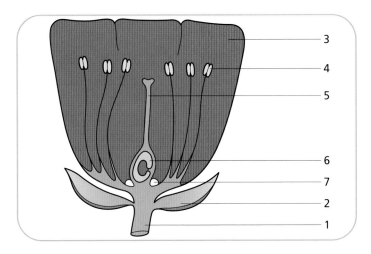

Figure 3.3 Inside a simple flower

Part	Function
1 Flower stalk	Supports flower.
2 Sepal	Protects flower bud before opening.
3 Petal	Attracts insects and protects other flower parts.
4 Anther	Produces pollen, the male sex cells. Together with its stalk, known as a stamen.
5 Stigma	Sticky platform onto which insects deposit pollen from other flowers.
6 Ovary	Produces ovules, the female sex cells. This part later forms the fruit and seeds.
7 Nectary	Produces sugary nectar to attract insects.

Pollination

G4 Describe methods of pollination.

Insect pollination (brightly coloured, scented flowers with nectar – features which attract insects)

An insect enters one flower in search of nectar and brushes against the ripe anthers. Pollen sticks to the insect's body. The insect flies to another flower and if the stigma is ripe, the pollen from the first flower will stick to it, so pollinating the flower.

Wind pollination (small dull flowers with no nectar, e.g. grasses)

Large numbers of light pollen grains are produced by large anthers hanging down outside the flower. The pollen grains are carried away in the wind. Some may be caught on the large feathery stigma hanging out of another flower, so pollinating the flower.

Insect- and wind-pollinated flowers compared

C2 Explain the structure of wind- and insect-pollinated flowers in relation to sexual reproduction.

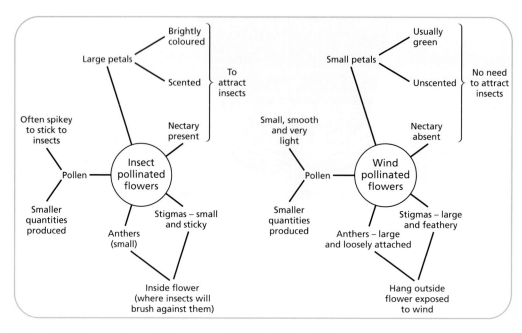

Figure 3.4 **Comparison of insect- and wind-pollinated flowers**

Fertilisation and fruit formation

G5 Describe fertilisation and fruit formation.

C3 Describe the growth of the pollen tube and the fusion of gametes.

After pollination (the transfer of pollen from one flower to another by insects or wind), the male gamete inside the pollen grain has to travel to the ovary and fuse (join) with the female gamete.

HOW TO PASS STANDARD GRADE BIOLOGY

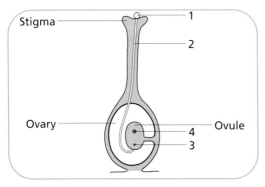

Figure 3.5 Fertilisation of an ovule

The pollen grain (1) grows a pollen tube (2) which grows down from the stigma towards the ovary. The tip of the tube enters the ovule and the male gamete (3) which has been carried along in the tube enters the ovule and fuses with the female gamete. (4) The joining of the two gametes is called fertilisation.

After fertilisation, the fruit develops as shown below.

Table 3.2 Development of fruit after fertilisation

Flower part		Fruit part
Ovule(s) (inner part)	→	Embryo and food store (seed)
Ovule(s) (outer part)	→	Seed coat
Ovary wall	→	'Fleshy' part (e.g. tomato) or hard outer case (e.g. hazelnut)

Seed dispersal

C4 Describe one example of each of the following different seed dispersal mechanisms (wind, animal – internal, animal – external).

After fertilisation, the ovule develops into a seed containing an embryo plant.

Seeds must be dispersed (spread) away from the parent plant to avoid overcrowding and competition between parent and seedlings.

Three mechanisms of seed dispersal are shown in Table 3.3.

HOW TO PASS STANDARD GRADE BIOLOGY

Table 3.3 Seed dispersal mechanisms

Method	Example	Explanation
Wind Figure 3.6	Dandelion Lime Sycamore Seed	Light seeds with wing-like extensions or 'parachutes' blown some distance from the parent plant
Animal (internal) Figure 3.7	Cherry Raspberry Seed Seed	Seeds contained in a fruit (usually soft and juicy). Animals eat the fruit, but seeds pass out undigested in droppings some distance from the parent plant
Animal (external) Figure 3.8	Agrimony	Dry hooked fruits containing seeds catch on passing animals and drop off later, some distance from the parent plant

Artificial propagation

G6 **Describe ways of propagating flowering plants artificially by cuttings and grafting.**

Cuttings

1 The stem is cut just below a leaf or leaf bud.

2 The cut stem is dusted with hormone powder to encourage growth of new roots.

3 The cutting is potted up in damp compost.

4 After a few weeks, new roots develop and the cutting is now an independent plant.

Grafting

1 Part of a stem or branch of one plant (the scion) is fitted into a notch cut in the root stock of another plant.

2 The graft is bound together and after a few weeks, the scion and stock will have grown together to produce a single healthy plant.

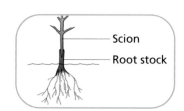

Figure 3.9 An example of grafting

Advantages of artificial propagation

C5 Explain the advantages to humans of artificial propagation in flowering plants.

1 Identical offspring: All the desirable features of the parent plant are retained in the offspring (a clone).

2 Speed: Many plants can quickly be produced from one parent (e.g. by taking cuttings).

3 Grafting: The best features of two different plants can be combined in one.

Cloning plants

C6 State what is meant by the term 'clone'.

Artificial propagation can produce many new plants all from the same parent. As the new plants have grown from pieces of the parent plant, they will all be identical to the parent plant and to each other. The identical offspring of a single parent produced in this way are known as **clones**.

Asexual reproduction in plants

G7 Describe asexual reproduction by runners and tubers.

Many plants can reproduce by the process of asexual reproduction.

Asexual = non-sexual i.e. not involving gamete cells from two parents.

Runners (e.g. strawberry)

A parent plant sends out a horizontal shoot. Where this touches the soil a new plant will form. The parent provides food through the runner until the new plant can make its own, then the runner shrivels away.

Tubers (e.g. potato)

A potato plant sends out underground stems which swell up at the tip to form potatoes. The parent plant dies in the autumn. The following spring, each potato sends out shoots and roots, becoming a complete plant and starts the process again.

HOW TO PASS STANDARD GRADE BIOLOGY

Advantages and disadvantages of sexual and asexual reproduction in plants

C C7 Describe the advantages and disadvantages of both sexual and asexual reproduction in plants.

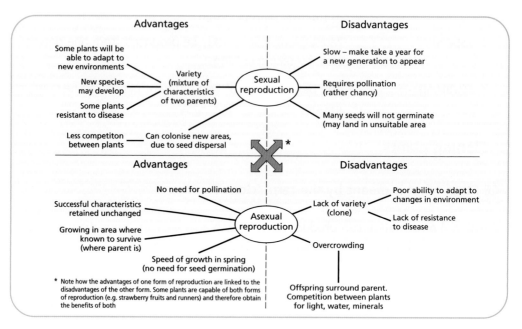

Figure 3.10 **Asexual and sexual reproduction in plants**

Questions

G

1 Explain why it is important that seed germination will not occur if the temperature is low.

2 Describe the part played by water in bringing about germination.

3 Construct a table to show the differences between insect- and wind-pollinated flowers.

4 Which part of a cherry flower develops into the juicy flesh of a cherry after fertilisation?

C

5 Describe how a male gamete inside a pollen grain reaches and fetilises a female gamete (ovule) inside the ovary of a flower.

Questions continued ➣

Questions *continued* **?**

6 Give examples of plants that use each of the following methods of seed dispersal: (a) wind; (b) animal (internal); (c) animal (external).

7 Describe one advantage and one disadvantage of **sexual** reproduction in plants.

8 Describe one advantage and one disadvantage of **asexual** reproduction in plants.

Making Food

Plant transport systems

 G1 Explain the need for transport systems in a plant.

Unlike animals, plants have two transport systems:

1 To transport water and minerals from the soil to all parts of the plant.

2 To transport food made in the green parts of the plant (mainly the leaves) to all non-food-making parts of the plant (e.g. roots).

 G2 Describe the movement of water through xylem and food through phloem.

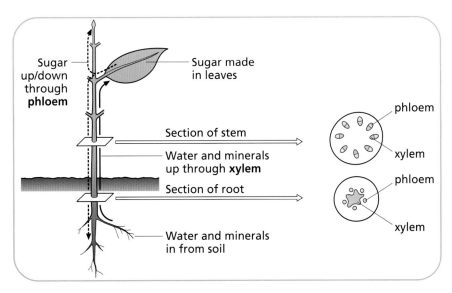

Figure 3.11 Movement of water and food through a plant

Xylem

Water is taken in through root hairs and transported up the stem through xylem vessels. These are hollow columns of dead cells and they form continuous tubes reaching all parts of the plant.

Phloem

Phloem tissue takes sugar solution from the leaves to non-food-producing parts of the plant.

Vascular bundles

1   Roots: xylem tissue forms a star-shaped rod of cells running up the middle of the root. Smaller rods of phloem tissue lie between the arms of the star (see Figure 3.11, p. 49).

2   Stem: vascular bundles lie around the outside of the stem. Each is a rod of phloem tissue to the outside and xylem tissue to the inside.

Veins

Veins are vascular bundles that leave the stem and go out into a leaf.

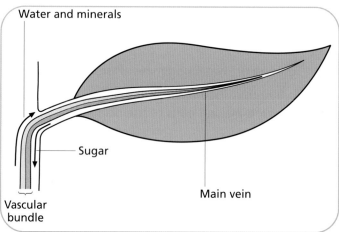

Figure 3.12 Veins in a plant

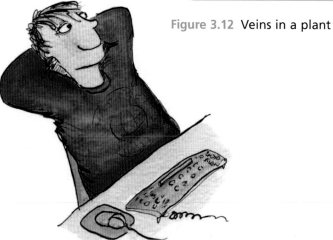

Structure of xylem and phloem

C1 Describe the structure of xylem and phloem and identify other functions of the transport system.

Xylem

Part of a xylem vessel showing bands of lignin. This tough woody material strengthens the xylem.

As well as transporting water and minerals, xylem strengthens roots and stems. Lignified xylem is what makes wood rigid.

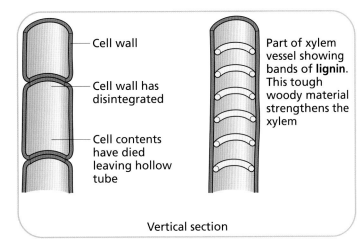

Cell wall

Cell wall has disintegrated

Cell contents have died leaving hollow tube

Part of xylem vessel showing bands of **lignin**. This tough woody material strengthens the xylem

Vertical section

Figure 3.13 Vertical section of xylem

Phloem

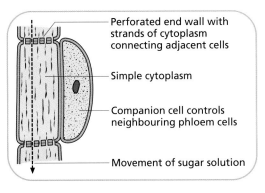

Perforated end wall with strands of cytoplasm connecting adjacent cells

Simple cytoplasm

Companion cell controls neighbouring phloem cells

Movement of sugar solution

Figure 3.14 Vertical section of phloem

Taking in carbon dioxide...

G3 State that plants take in carbon dioxide from the air through stomata which can open and close.

Fact: Carbon dioxide gas is needed by green plants as a raw material for making carbohydrate.

Fact: Carbon dioxide is present in air. (CO_2 makes up approximately 0.04% of atmospheric air.)

Problem: The outer surfaces of leaves are almost waterproof and gas proof. Carbon dioxide cannot easily get through them.

Solution: Leaves have many tiny holes called **stomata** (singular, stoma) which the plant can open and close. Carbon dioxide can enter the leaf through these stomata.

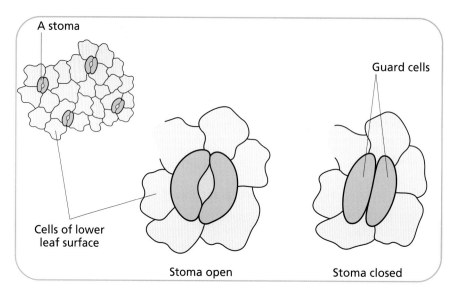

A stoma

Guard cells

Cells of lower leaf surface

Stoma open   Stoma closed

Figure 3.15 View of the lower surface of a leaf

...but losing water

G **G4 State that water vapour is lost through stomata.**

When the stomata are open to allow carbon dioxide to enter the leaf, water vapour diffuses out in the other direction. Too much evaporation of water vapour from the leaves may cause the cells to become flaccid and the plant wilts. When this happens, the guard cells close together, shutting the stomata and reducing water loss.

Structure of a typical leaf

C2 Describe the external features and internal structure (epidermis, mesophyll, veins) of a leaf in relation to its function in gas exchange.

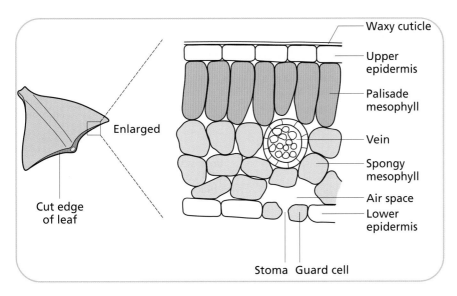

Figure 3.16 External and internal structure of a leaf

Table 3.4 Function of leaf structures

Structure	Function
Waxy cuticle	Reduces evaporation of water through the epidermis
Upper and lower epidermis	Protect more delicate cells inside
Palisade mesophyll	Where most photosynthesis takes place. These cells contain many chloroplasts
Spongy mesophyll	Some photosynthesis. Not as many chloroplasts as palisade cells
Air spaces	Allow gases to move freely around inside the leaf
Stoma	Allow gas exchange with the atmosphere (CO_2 in, O_2 and water vapour out)
Guard cells	Control opening and closing of stomata
Vein	Delivers water and minerals. Takes away sugars produced by mesophyll cells

Making starch

G5 State that green plants make their own food which may be stored in the form of starch.

G6 State that green leaves convert light energy to chemical energy using chlorophyll.

Plants make their own food by the process called **photosynthesis** (**photo** = with light; **synthesis** = making).

The end-product of photosynthesis is sugar which may be converted to starch for storage. This can be shown by the following experiment.

After several days the leaves are tested for starch.

Result: Leaf A – starch absent. Leaf B – starch present.

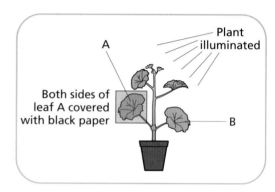

Figure 3.17 Proof of starch production

Conclusion: The plant has used light energy to make starch, i.e. it has converted light energy into chemical energy (in the starch).

The plant should be 'destarched' before starting, by being placed in the dark for several days, causing any starch in the leaves to be used up.

Starch test for leaves

1 Leaf boiled in water to kill and soften leaf.

2 Leaf boiled in alcohol to remove green colour (chlorophyll) to make starch test easier to see.

3 Leaf softened again in water.

4 Leaf covered in iodine: black = starch present, brown = starch absent.

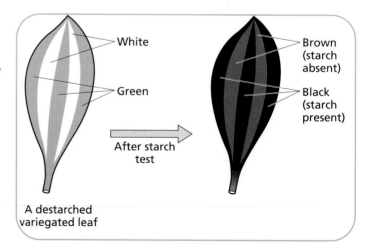

Figure 3.18 Starch test on a variegated leaf

That the green pigment chlorophyll is essential for photosynthesis can be shown by carrying out the starch test on a variegated leaf (one which is not all green).

Conclusion: Starch is only produced in areas where chlorophyll is present.

Carbohydrates in the plant

C3 Describe the fate of carbon dioxide as structural and storage carbohydrates in plants and as energy sources.

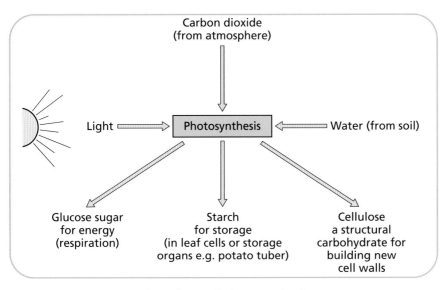

Figure 3.19 Summary of products of photosynthesis

Summary of photosynthesis

G7 Describe the process of photosynthesis in terms of raw materials and products.

Table 3.5 Process of photosynthesis

Raw materials	Energy source	Products
Carbon dioxide + water	Sunlight (chlorophyll also needed to absorb light, but is not used up)	Sugar + oxygen (sugar stored as starch)

C

Limiting factors

C4 Explain what is meant by a limiting factor and describe the main limiting factors in the process of photosynthesis.

Environmental factors all needed for photosynthesis:

1   sufficient heat
2   light
3   carbon dioxide.

If any of these is lacking, photosynthesis will not take place as fast as it could.

1   Frosty but sunny winter day: temperature is too low for enzymes to work well, so temperature is the limiting factor.
2   Cloudy but warm summer day: light intensity is below optimum, so light is the limiting factor.
3   Warm sunny summer day: carbon dioxide is now likely to be the limiting factor.

G

C

Questions

1   Describe how water moves from the soil into the leaves of a plant.

2   Describe how carbon dioxide passes from the atmosphere into a leaf.

3   Write a word equation describing the process of photosynthesis.

4   Name the plant transport tissue responsible for each of the following functions:

(a) transport of water;

(b) transport of sugar;

(c) support of the stem.

5   Name the internal leaf layer where most chloroplasts are found.

6   Describe three ways a plant may use carbohydrate produced by photosynthesis.

7   Explain how temperature and light may be limiting factors in photosynthesis.

Chapter 4

ANIMAL SURVIVAL

> ### The Need for Food
>
> **Why do we need to eat?**

G1 Explain in simple terms why food is required by animals.

Animals, including humans, need to eat, because food provides the chemicals needed for:

1. energy (see aerobic respiration, p. 36)
2. growth of new cells
3. repair of damaged cells.

Food also provides essential vitamins, minerals and water, without which cells would not operate correctly.

The main food types

C1 State the chemical elements present in carbohydrates, proteins and fats.

C2 Describe the structure of carbohydrates, proteins and fats in terms of simple sugars, amino acids, fatty acids and glycerol.

Carbohydrates

Examples: sugars (e.g. glucose, sucrose) and starch.

Function: energy giving foods.

Chemical elements: carbon hydrogen oxygen.

Glucose   single simple sugars molecules

Maltose   pairs of simple sugar molecules

Starch   ... long chains of simple sugars

Figure 4.1 **Examples of carbohydrates**

carbo — hydr — ate   (**ate** ending means 'containing oxygen')

Structure: All carbohydrates are made up of one or more **simple sugars**.

Proteins

Examples: lean meat, cheese, egg white, nuts.

Function: needed for growth and repair of cells.

Chemical elements: carbon, hydrogen, oxygen and nitrogen.

Structure: all proteins are made up of long chains of smaller molecules called **amino acids.**

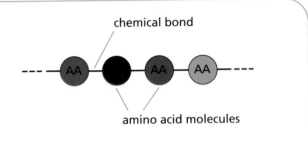

About 20 different types of amino acids can be joined together in different combinations and shapes, producing many different proteins.

Figure 4.2 Protein structure

Fats

Examples: milk products, margarine, animal fats, oily fish (e.g. herring).

Function: energy store, heat insulation.

Chemical elements: carbon, hydrogen, oxygen.

Structure: fats are made up of two different types of smaller molecules.

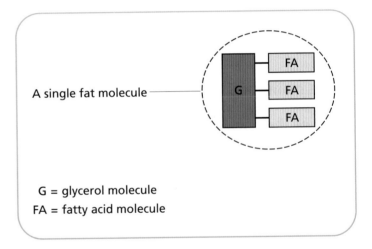

G = glycerol molecule
FA = fatty acid molecule

Figure 4.3 Structure of a single fat molecule

The need for digestion

G2 State that digestion is the breakdown of large particles of food into smaller particles to allow absorption through the wall of the small intestine and into the blood stream.

1   The wall of the small intestine is made of cells.

2   Food must pass through these cells before it can enter the blood capillaries surrounding the intestine.

3   Only small molecules can pass through cell membranes (see Chapter 2, p. 24).

4   Most food is made up of very large molecules (e.g. starch, proteins, fats etc.).

5   Food must therefore be broken down (digested) into molecules small enough to pass through the gut wall.

Insoluble to soluble food

C3 Explain that digestion involves the breakdown of insoluble food substances into soluble food substances.

Only liquids can pass through the gut wall, so food must be dissolved in water before it can pass through. Not only are large molecules such as starch too big to pass through the membrane, they are also virtually insoluble. Digestion turns them into smaller, soluble molecules so that they may pass through the membranes of the gut lining.

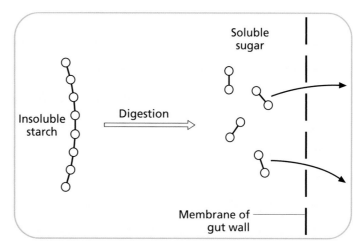

Figure 4.4 Breakdown of insoluble to soluble food substances

The action of teeth

G3 Describe the role of different types of teeth in the mechanical breakdown of food in: a human (omnivore); a carnivore (e.g. dog); and a herbivore (e.g. sheep)

Human teeth

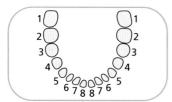

Figure 4.5 Human teeth

1, 2 and 3 = molars
4 and 5   = premolars
6   = canines
7 and 8   = incisors

The upper and lower jaw are identical in humans.

Table 4.1 A human

Tooth type	Shape	Function
Incisors	Chisel shaped	Cutting and biting
Canines	Sharp and pointed	Biting and tearing
Premolars	Flattened, but with two cusps (raised parts)	Tearing and grinding
Molars	Large and flattened, but with four or five cusps	Chewing and grinding

Table 4.2 A carnivore (e.g. dog)

Tooth type	Shape	Function
Incisors	Sharp, pointed	Biting off small pieces of meat
Canines	Large, sharp and pointed	Gripping and killing prey
Premolars and molars	Scissor-like or flattened	Slicing meat Crushing bones

Table 4.3 Herbivore (e.g. sheep)

Tooth type	Shape	Function
Incisors and canines	Small and chisel shaped	Gripping and tearing vegetation
Premolars and molars	Large, broad and with enamel ridges	Grinding and chewing vegetation

The digestive system of mammals

G4 Identify the main parts of the mammalian alimentary canal and associated organs.

You should be able to name all of the following parts of the mammalian digestive system (alimentary canal):

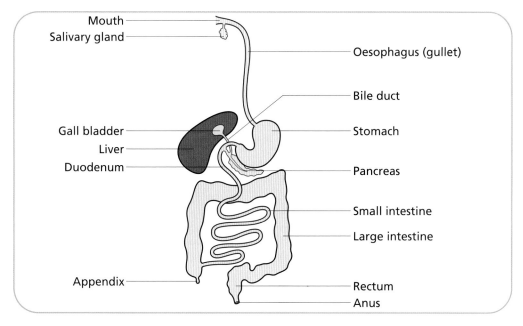

Figure 4.6 Mammalian digestive system

Production of digestive 'juices'

C4 State the sites of production of the main digestive juices in a mammal.

Table 4.4 Digestive juices

Digestive juice	Where produced	Substrate	Produce
Saliva	Salivary glands in mouth and tongue	Starch	Maltose
Gastric juice	Stomach lining	Protein	Peptides
Pancreatic juice	Pancreas	Proteins and peptides Starch Fats	Amino acids Maltose Fatty acids and glycerol
Intestinal juice	Lining of small intestine	Peptides Fats Maltose	Amno acids Fatty acids and glycerol Glucose
Bile	Liver	Bile is not a digestive enzyme. It emulsifies fat, i.e. it breaks large lumps of fat into many small lumps, to increase the surface area to speed up enzyme action.	

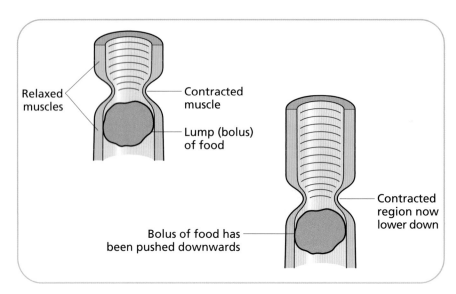

Figure 4.7 Movement of food through the digestive system (Peristalsis)

Swallowing…and further

C5 Explain the mechanism of peristalsis.

Food is pushed through the digestive system by muscles surrounding the gut. Waves of muscular contraction called **peristalsis** run from the top to the bottom of the alimentary canal, keeping the food moving.

C6 Explain how the contractions of the stomach help in the chemical breakdown of food.

Muscles in the stomach wall churn the food and digestive juices together. This mixing helps the digestive juices to begin the chemical breakdown of food (particularly protein).

Enzyme action

G5 State that different enzymes are responsible for the breakdown of carbohydrates, proteins and fats.

As food passes through the alimentary canal, enzymes are produced which break the food down into the smaller molecules of which food is made, e.g. stomach lining produces protein-digesting enzymes, pancreas produces fat-, protein- and carbohydrate-digesting enzymes.

Some digestive enzymes

C7 Give an example of an amylase, a protease and a lipase. State their substrates and products.

Table 4.5 Digestive enzymes

Name of enzyme	Where produced	Substrate	Product
Salivary amylase	Salivary glands	Starch	Maltose
Pepsin (a protease or protein-digesting enzyme)	Stomach lining	Protein	Peptides
Pancreatic lipase	Pancreas	Fats	Fatty acids and glycerol

Absorbing the products of digestion

G6 Explain how the structure of the small intestine is related to its function.

The structure of the small intestine is very efficient at absorbing digested food for the following reasons:

1   The small intestine is **very long.**

2   The inner surface is folded into millions of finger-like **villi** (singular, villus).

These two features provide a **very large surface area** for absorption.

3   The lining of the villi is **very thin**, allowing digested food to pass quickly into the blood stream.

Absorption of food in the small intestine

C8 Explain how the structure of a villus including the lacteal and the blood capillaries are related to the absorption and transport of food.

The surface area of the small intestine is greatly increased by millions of tiny hair-like structures called **villi**. This increased surface area is necessary for efficient absorption of digested food.

Glucose (digested carbohydrate) and amino acids (digested protein) pass through the outer layer of cells and into blood capillaries which carry them away to the rest of the body.

Fatty acids and glycerol (digested fat) enter the central lacteal. From there, they are transported by the lymphatic system, eventually being emptied into the bloodstream elsewhere in the body.

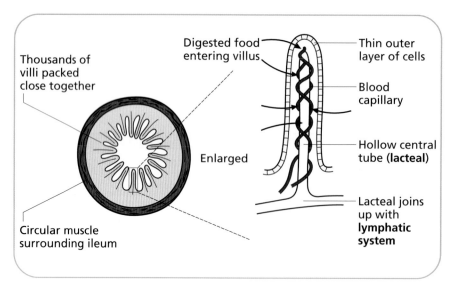

Figure 4.8 A single villus

The large intestine

(**G**) **G7** Describe the role of the large intestine in water absorption and elimination.

Undigested food, largely fibre (i.e. cellulose from plant material), passes into the large intestine along with a lot of water.

The main function of the large intestine is to absorb much of this water to prevent the body from drying out. Water passes through the walls of the large intestine into the blood stream and the semi-solid waste that remains is stored in the rectum until it passes out through the anus. This is called **elimination** and the solid wastes are called **faeces**.

Questions (?)

1   Describe three reasons why food is required by animals.

2   Rearrange the following parts to show the sequence taken by food passing down the alimentary canal:
     small intestine; rectum; stomach; anus; large intestine; oesophagus.

3   Name the finger-like projections that give the small intestine a very large surface area.

Questions continued ➢

Questions *continued*

4 Construct a table for carbohydrates, proteins and fats, showing:

(a) chemical elements present;

(b) sub-units making up the complete molecules;

(c) function(s) in the body.

5 Describe how peristalsis pushes food through the alimentary canal.

6 Name a protease (protein-digesting enzyme) and state its substrate and products.

Reproduction

Eggs and sperm

G1 Describe the main features of sperm and eggs.

Table 4.6 Comparison of eggs and sperm

Eggs	Sperm
Produced by females	Produced by males
Smaller numbers produced	Very large numbers produced
Much larger than sperm	Much smaller than eggs
Are not 'self propelled'	Have 'tails', so can swim
Contain food reserves	Have very limited food reserves

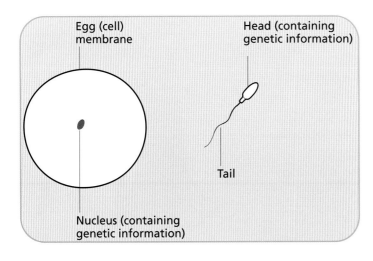

Figure 4.9 Features of egg and sperm cells

External or internal fertilisation

(G) **G2** State that in some fish, sperm are deposited in water close to the eggs and that in mammals, sperm are deposited in the body of the female.

(G) **G3** Describe the process of fertilisation.

Sexual reproduction involves the joining together (fusion) of a reproductive cell (a gamete) from a male, with another from a female.

Males produce sperm cells and females produce egg cells (sometimes called ova).

When a sperm and an egg fuse together, the nuclei join and their genetic information 'mixes' together. This process is called **fertilisation** and leads to a new individual having a mixture of characteristics from both parents.

How do sperm and eggs meet?

In many animals, including most fish, the female lays her eggs into the water and the male releases his sperm near the eggs. The sperm then swim through the water to reach the eggs. This is **external fertilisation** as it takes place outside the animal's body.

In other animals, including all mammals, the female retains the egg in her body and the male deposits sperm directly into the female's body. **Internal fertilisation** then takes place inside the female's body.

Fertilisation in land-living animals

(C) **C1 Explain the importance of internal fertilisation to land-living animals.**

Land-living animals do not have water into which they can release eggs and sperm. The eggs, therefore, are retained inside the female's body and during the act of copulation, sperm are passed by the male into the female's body. Once inside, they are in a watery environment and can swim towards the egg.

Production of eggs and sperm

G4 State that sperm are produced in testes.

G5 State that eggs are produced in ovaries and are released into oviducts, where, in mammals, fertilisation takes place.

In all male animals, sperm are made in testes.

In all female animals, eggs are made in ovaries.

Fertilisation in mammals

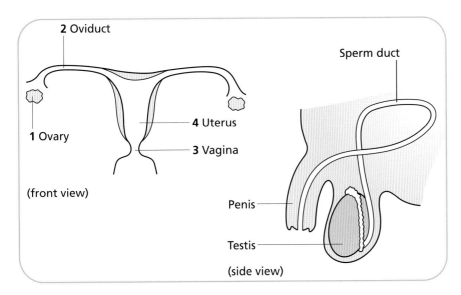

Figure 4.10 **Female and male reproductive organs**

1 The ovary releases an egg.
2 The egg passes down the oviduct.
3 If copulation takes place, the male's penis releases sperm into the vagina.
4 Sperm swim up through the uterus and if they meet the egg coming down the oviduct, one sperm will fuse with the egg and fertilisation will take place.

Development in fish

G6 State that in fish, eggs are protected by flexible coverings and that the embryos obtain food from the enclosed yolk.

G7 State that in a fish, such as trout, the young emerge from the eggs able to maintain themselves.

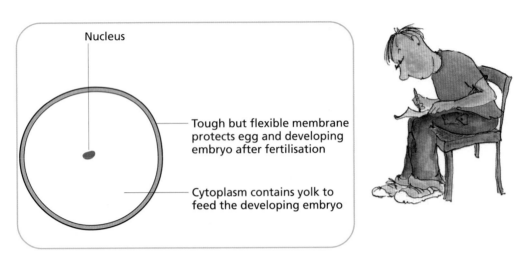

Nucleus

Tough but flexible membrane protects egg and developing embryo after fertilisation

Cytoplasm contains yolk to feed the developing embryo

Figure 4.11 Features of a fish egg

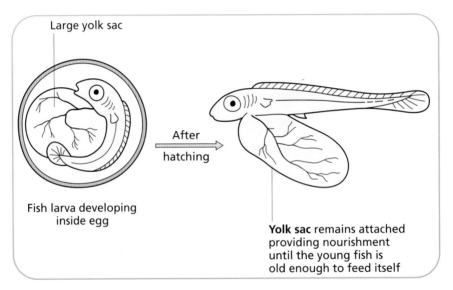

Large yolk sac

After hatching

Fish larva developing inside egg

Yolk sac remains attached providing nourishment until the young fish is old enough to feed itself

Figure 4.12 Development of a fish egg

Improving the chances of successful reproduction

C **C2 Explain the relationship between the number of eggs/young produced and the amount of protection given during fertilisation and development in fish and mammals.**

Table 4.7 Fish and mammal eggs

	Number of eggs	Chances of fertilisation	Protection of eggs	Parental protection of young	Chances of successful development
Fish	Large number	Low	None (or very little)	None (or very little)	Poor
Mammals	Few	High	Protected inside mother's body	Young are dependant on parents for food and protection	High

Development of mammals

G8 Describe how the fertilised egg passes down the oviduct and becomes attached to the wall of the uterus, develops in fluid of the amniotic sac and obtains food from the maternal circulation.

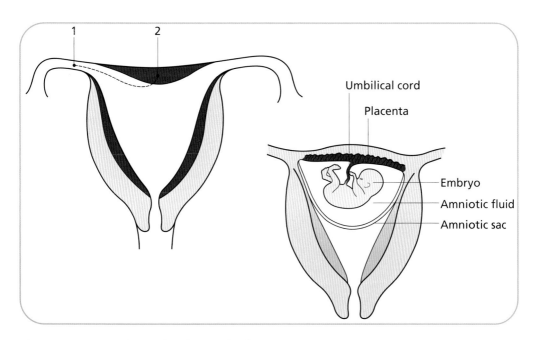

Figure 4.13 An embryo developing in the uterus

1   After fertilisation in the oviduct, the egg divides rapidly.
2   When it arrives in the uterus, it is a ball of cells and it sinks into the wall of the uterus. From now on, it receives food and oxygen from the mother's blood as it continues to develop.

The embryo's blood is pumped through the umbilical cord to the placenta where it picks up food and oxygen.

The embryo is bathed and protected by amniotic fluid contained in the amniotic sac.

Structure and function of the placenta

C C3 **Describe the structure and function of the placenta.**

Functions of the placenta
To allow exchange of substances between the mother's and the embryo's blood.

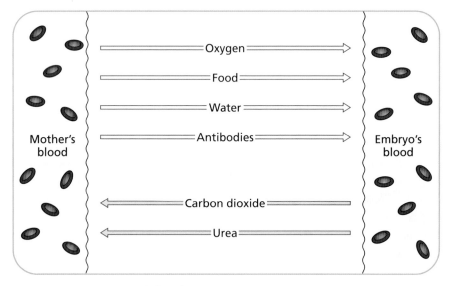

Figure 4.14 Function of the placenta

These exchanges all take place without the mother's and embryo's blood coming in contact. This is important, as they may have different blood types.

Structure of the placenta

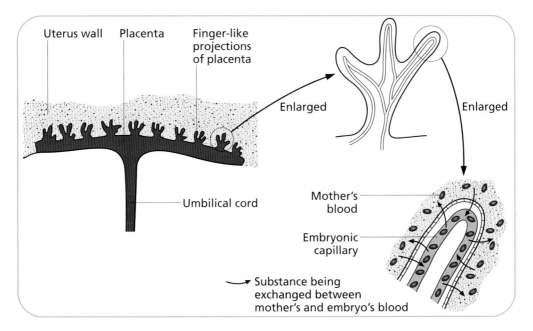

Figure 4.15 Structure of the placenta

Care after birth

G9 State that at birth, the young of mammals are dependant on the adult for care and protection.

Unlike fish, mammals depend on their parents for some time after birth. As well as feeding the young mammals on milk produced by the mother, the parents protect the young from danger and cold.

Questions

1 Construct a table showing the main features of eggs and sperm.

2 Describe the differences between internal and external fertilisation in animals.

3 In mammals, where is fertilisation of an egg most likely to occur?

4 Describe the function of the placenta.

⇨ ## *Water and Waste*

Water gain and loss in animals

G **G1 Identify the ways in which a mammal gains and loses water.**

Table 4.8 Water gain and loss

Water gain	Water loss
Food: most food contains a lot of water, especially fruit, vegetables, meat etc.	**Urine:** liquid waste, mostly a solution of **urea** in water
Drink: all drinks have a high water content.	**Faeces:** semi-solid waste, also contains a lot of water
Body chemistry: chemical reactions in the body make a small amount of water.	**Sweat:** more lost in hot weather, but we lose some water through sweating even in cold weather
	Breath: exhaled air comes from the lungs saturated with water vapour

Structure and function of the kidneys

G **G2 State that the kidneys are the main organs for regulating the water content in a mammal.**

G **G3 Identify the position and state the functions of: the kidney, renal arteries and veins, ureter and bladder.**

Water intake will sometimes be more, sometimes less than water loss. Regulating the amount of water in the body, so that a balance is maintained, is achieved by the kidneys.

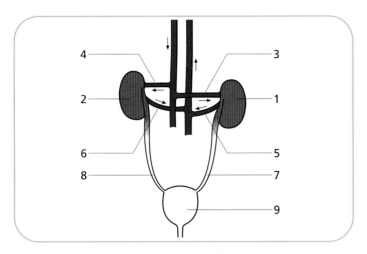

Figure 4.16 The kidneys and their blood vessels

Name	Function	
1 & 2	Left and right kidneys	Filter blood. Remove excess water and urea.
3 & 4	Renal arteries	Supply blood to the kidneys.
5 & 6	Renal veins	Take filtered blood away from the kidneys and back into the general circulation.
7 & 8	Ureters	Take urine (urea and water) from the kidneys to the bladder.
9	Bladder	Store urine until it can be passed out of the body.

How the kidneys work

G4 State that the kidneys work by filtration of blood and re-absorption of useful materials such as glucose.

G5 State that urea is a waste product removed in the urine.

Filtration
As blood flows through the kidneys, a lot of the water and most of the substances dissolved in the plasma are filtered out. This includes urea, glucose, amino acids and salt.

Re-absorption
Many of these substances are needed by the body, so they are then taken back into the blood. This is called re-absorption. Urea (a waste product) and just the right amount of water that the body needs to lose are not re-absorbed and so pass out of the kidneys, down to the bladder and out of the body as urine.

Urine production in the kidneys

C

C1 Explain the process of urine production using a simple diagram of the nephron.

Each kidney has approximately one million nephrons.

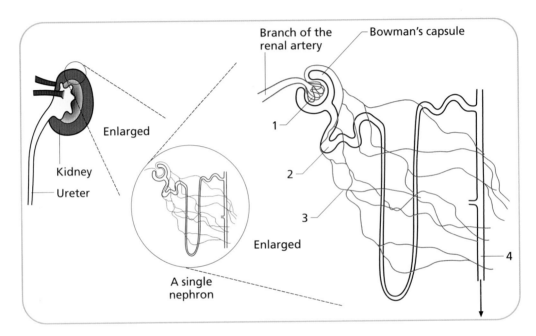

Figure 4.17 Urine production in the kidneys

Part of nephron	Function
1 Capillaries (glomerulus)	Blood is filtered and collects in the capsule.
2 Kidney tubule	Useful substances re-absorbed (e.g. glucose, amino acids) and pass back into the capillaries.
3 Capillaries	
4 Collecting duct	Final re-absorption of water. The amount re-absorbed here is controlled by anti-diuretic hormone (ADH).

The source of urea

C

C2 State the source of urea in the body and describe how urea is transported to the kidneys.

Urea is formed in the liver from the breakdown of surplus amino acids.

Amino acids → (enzyme controlled breakdown) → **glycogen** (a useful carbohydrate, stored in the liver) + **urea** (a poisonous nitrogen compound).

The urea then passes into the bloodstream and is carried round the body until it reaches the kidneys where it is removed from the blood.

Regulation of water balance

C3 Explain the role of ADH in the regulation of water balance.

The amount of anti-diuretic hormone (ADH) secreted by the brain depends on the concentration of the blood.

If the blood is too watery (e.g. after drinking a lot) ADH production is reduced and the kidneys produce lots of dilute urine.

If the blood is too concentrated (e.g. when dehydrated or after eating a lot of salt) ADH production is increased and the kidneys produce a small amount of concentrated urine.

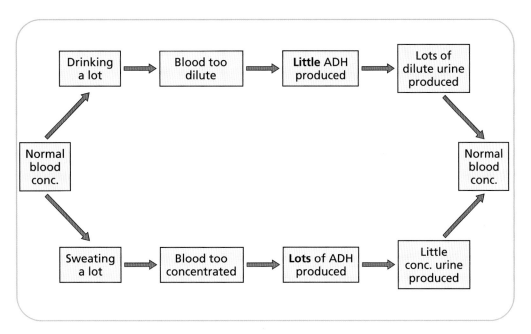

Figure 4.18 Regulation of water balance

Kidney disease

G6 Explain the implications of damage to the kidneys by accidents or disease.

If both kidneys stop working, either through disease or an accident, the body will be unable to:

1   remove poisonous urea from the blood

2   regulate the amount of water in the body and excess water will build up in the body.

Through a combination of these two, total kidney failure quickly causes death, unless treated.

When kidneys stop working

C C4 **Describe the benefits and limitations of replacement and 'artificial' kidneys.**

A kidney failure can be treated by:

1   A kidney transplant – a donor kidney from another person is placed inside the patient's body and connected to an artery and to the bladder.

2   Kidney machines – several times a week the patient's blood is passed through a machine which removes urea and excess water.

Each method has advantages and limitations, as shown in the table below.

Table 4.9 Treatment for kidney failure

	Benefits	Limitations
Transplant	◆ Patient can live an almost normal life (e.g. can go on holiday). ◆ Not dependant on a machine. ◆ Patients can eat and drink normally.	◆ Tissue rejection may cause new kidneys to fail. ◆ Patient on drugs for rest of life. ◆ Drugs may make patients more likely to catch infectious diseases. ◆ Not enough suitable kidneys available.
Kidney machine	◆ No drugs necessary. ◆ Availability of machines only limited by money	◆ Patient's diet very restricted. ◆ Patient must always be near a machine. ◆ Machines may break down.

Questions

1   As well as from food and drink, how else does a mammal gain water?

2   Briefly describe the two main functions of the kidneys.

3   Name two substances re-absorbed back into the bloodstream after filtration of the blood in the kidneys.

4   What is the full name of the hormone normally abbreviated to ADH?

5   Choose the correct alternatives to complete the following sentence:

When blood is too concentrated due to sweating/drinking, more/less ADH is produced, leading to a larger/smaller volume of dilute/concentrated urine.

6   After kidney failure, explain why a kidney transplant is regarded as more desirable than using an 'artificial' kidney.

Responding to the Environment

Environmental factors that affect behaviour

G1 Give examples of environmental factors that affect behaviour.

C1 Explain the significance of responding to environmental stimuli.

		Importance
1	Frogs remain in damp areas	They need water for external fertilisation. Also, they need moist skin to assist in gas exchange.
2	Fish avoid water low in oxygen	If they swim into water with little oxygen, they may suffocate.
3	Birds begin nesting in response to longer daylength in spring.	Young will have a plentiful supply of food in the warmer weather of late spring/early summer.

Animal responses to changes in environmental factors

G2 Describe the response of an animal to change in one environmental factor.

The response of woodlice to light and dark can be shown using a choice chamber.

Woodlice do not move deliberately towards the dark area. They move at random within the choice chamber and when they enter the dark area, they slow down or stop, thus remaining for longer in a more suitable area.

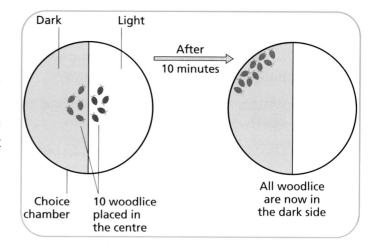

Figure 4.19 Animal responses to changes in environmental factors

Rhythmical behaviour in animals

G3 Describe examples of rhythmical behaviour and in each case identify the external trigger stimulus.

Table 4.11 Rhythmical behaviour in animals

Examples of rhythmical behaviour	Trigger stimulus
Bird migration	Changes in daylength
Hibernation of bears	Change in environmental temperature
Cockroaches more active at night	On-set of darkness
Mating of Grunion (a fish)	High tide (caused by position of the moon)

The trigger stimulus is the particular environmental factor which causes the behaviour to start.

The importance of rhythmical behaviour in animals

C2 Explain the significance of rhythmical behaviour in named animals.

Some examples from G3, p. 78:

		Importance
1	Geese migrate from the arctic circle to Europe in autumn (and back again in spring).	Avoid severe weather of an arctic winter. More food available in Europe.
2	Grunion gather for mating at a particularly high tide.	Ensures fertilisation of eggs by bringing males and females together.
3	Cockroaches become more active at night.	They avoid predators by feeding at night.

Questions ?

1   What environmental factor triggers migration in geese?

2   Give two more examples of rhythmical behaviour in animals and, in each case, identify the trigger stimulus.

3   Using two examples from questions 1 and 2, explain the benefit to the animals of this rhythmical behaviour.

THE BODY IN ACTION

 ## *Movement*

The skeleton – support, movement and protection

G **G1** State that the skeleton provides a framework for support and muscle attachment.

G **G2** State that the skeleton protects the heart, lungs, brain and spinal cord.

The skeleton has three main functions:

1   support: to hold us up against the force of gravity

2   movement: the skeleton provides a rigid framework for attachment of muscles

3   protection: all of the vital organs upon which the body depends for survival are protected by the skeleton.

Table 5.1 **Protection of vital organs**

Vital organ	Protected by
Brain	Skull
Spinal cord	Vertebrae of the backbone
Heart	Ribcage
Lungs	Ribcage

Joints

G **G3** Describe the range of movements allowed by a ball and socket joint and by a hinge joint.

Joints between neighbouring bones allow movement, but at the same time are strong enough to provide support. Most joints have a limited range of movement.

Table 5.2 **Body joints**

Type of joint	Examples	Movement allowed
Ball and socket	Hip, shoulder	Three planes of movement
Hinge	Knee, elbow	One plane of movement

Ligaments and cartilage

G4 State the functions of ligaments and cartilage at a joint.

In a joint, neighbouring bones must be allowed enough movement for the joint to work properly. They must not be allowed to move too much, however, or the joint would become dislocated.

Ligaments are tough fibrous tissues which hold bones together at a joint and prevent dislocation.

As a joint moves, neighbouring bones rub against one another. To reduce friction, the ends of the bones are covered with **cartilage** which is very smooth and slippery.

Cartilage is also rubbery, so it acts as a shock absorber when the bones are forced together.

Synovial joints

C1 Describe the structure of a synovial joint and state the functions of its parts.

A synovial joint is a joint lubricated by synovial fluid which is secreted by the synovial membrane surrounding the joint. The hip, shoulder, elbow, knee and finger are all synovial joints.

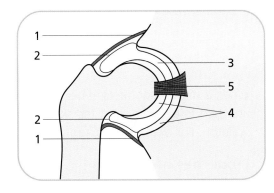

Figure 5.1 **Structure of a synovial joint**

Function of parts

1 **Capsule:** surrounds and protects the joint.

2 **Synovial membrane:** secretes synovial fluid.

3 **Synovial fluid:** sticky liquid that lubricates the joint.

4 **Cartilage:** smooth rubbery substance covering ends of bones to cushion the joint.

5 **Ligament:** holds the two bones together.

The structure of bone

G

G5 State that bone is composed of flexible fibres and hard minerals.

Healthy bones are hard and slightly bendy. The hardness provides support while the slight flexibility reduces breaks by allowing bones to bend slightly under stress.

The flexibility is provided by many **flexible fibres** running throughout the bone and hardness is provided by **minerals**.

C

C2 State that bone is formed by living cells.

1   Canal containing blood vessels.

2   Living cells which secrete calcium salts.

3   Hard bone made of calcium salts (mainly calcium phosphate).

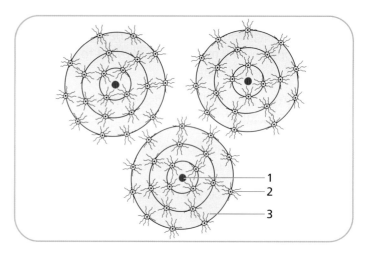

Figure 5.2 **Magnified section through bone**

Tendons

G

G6 State that muscles are attached to bones by tendons.

G

G7 Describe how movement is brought about by muscle contraction.

When a muscle contracts, it becomes shorter and fatter. One end is usually attached to a rigid part of the skeleton and the other to a moveable part of the skeleton. For example, raising the forearm (see Figure 5.3).

Properties of tendons

C

C3 Explain why tendons are inelastic.

As tendons are tough and inelastic (not stretchy) movements can be very fast and precise. The force created by the contraction of a muscle is transferred directly to the bone, making it move.

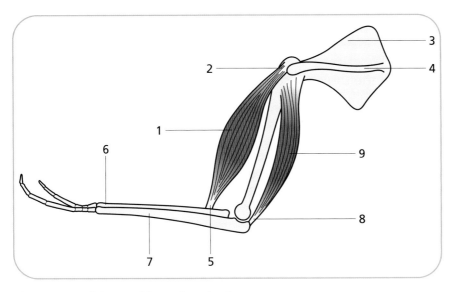

Figure 5.3 Raising and lowering the forearm

The **biceps** muscle (1) contracts. **Tendons** (2) pull on the **shoulder blade** (3) and **collar bone** (4) which cannot move, being attached to the rest of the skeleton. The **lower tendon** (5) pulls on the **radius** (6) which is raised. The **ulna** (7) is also raised and this pulls the tendon (8) which stretches the **triceps** muscle (9).

Opposing muscles

C4 Explain the need for a pair of opposing muscles at a joint.

Wherever a moveable joint is found, a pair of muscles operate, one moving the joint in one direction, the other muscle moving the joint in the other direction. These are called **opposing muscles**.

Questions

1   What are the three functions of the human skeleton?

2   Describe the differences in range of movement allowed by ball and socket and hinge joints.

3   Copy the following sentences, choosing correct alternatives:
Joints are held together by ligaments/cartilage. Cushioning of joints is provided by ligaments/cartilage. Muscles are attached to bones by ligaments/tendons.

4   (a) What is the function of synovial fluid in a joint?

     (b) Name a synovial joint.

5   Explain why it is important that tendons are inelastic (not stretchy).

6   Explain what is meant by the term 'opposing muscles'.

The Need for Energy

Energy sources

Muscular contraction causing body movement is one of the main uses of energy in the body.

There is a constant need for energy, as muscles are working all the time, awake or asleep (e.g. heart muscle).

The energy for this activity comes from food, in particular carbohydrates such as sugar and starch.

The amount of energy needed depends on age, sex and occupation.

Generally, children who are still growing need more energy than old people. Males need more energy than females.

Those with strenuous occupations (e.g. labourer) need more energy than those with less strenuous jobs (e.g. office worker).

The effect of energy imbalance

G1 State the effects of the imbalance between energy input and output.

Energy input = total energy of all food eaten in a certain time.

Energy output = total energy used by the body (for movement, making heat, etc.) in a certain time.

If energy input is more than energy output, the body will store the extra energy in the form of fat, and the body will gain weight.

If energy input is less than energy output, the body will take the extra energy it needs from stored body fat and the body will lose weight.

Exchange of gases

G2 State that oxygen is absorbed and carbon dioxide released in breathing.

Oxygen for aerobic respiration is obtained from the atmosphere when we take air into our lungs. Carbon dioxide produced as a waste product by aerobic respiration passes out of our body when we breathe out.

Table 5.3 **Exchange of gases when breathing**

	Composition of air	
	Inhaled (%)	Exhaled (%)
Oxygen	21	17
Carbon dioxide	0.03	4
Nitrogen	79	79

Structure of the lungs

G **G3 Describe the internal structure of the lungs.**

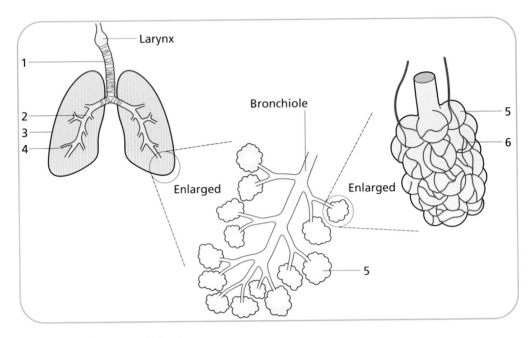

Figure 5.4 Structure of the lungs

Air passes from the mouth and nose down the **trachea** (1). This divides to form two **bronchi** (singular, bronchus) (2), one going to each **lung** (3).
Each bronchus divides many times into smaller **bronchioles** (4) which take air deep into the lung.
Each bronchiole ends in a number of thin-walled air sacs (5).
The air sacs are covered with a network of **blood capillaries** (6).

Mechanism of breathing

C **C1 Describe the mechanism of breathing in humans.**

Air is forced in and out of the lungs by contraction of the **intercostal muscles** and the **diaphragm muscles**.

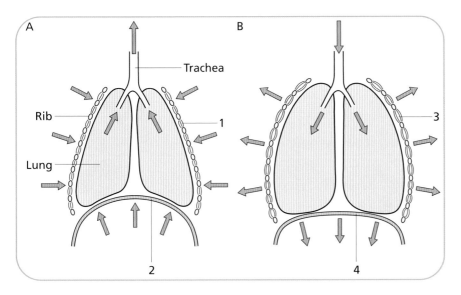

Figure 5.5 Features of the breathing mechanism

In Figure A, the intercostal muscles (1) and the diaphragm (2) are relaxed and in this position air is squeezed out of the lungs.

In Figure B, the intercostal muscles have contracted (3) lifting the ribcage upwards and outwards. The diaphragm muscles have pulled the diaphragm downwards (4). These two movements expand the lungs and draw in air.

Trapping dust and germs

C2 Explain the function of mucus, cilia and cartilage in the trachea and bronchi.

Dust and germs breathed in are trapped in sticky **mucus** produced by cells in the lining of the trachea and bronchi.

Tiny hairs called **cilia** in the lining of the trachea and bronchi push the mucus with trapped dirt up to the throat from where it passes down the oesophagus into the stomach. Acid in the stomach kills the germs.

The trachea and bronchi are strengthened by bands of **cartilage** which stop the air passages from collapsing during breathing.

Efficient gas exchange

C3 Describe the features which make lungs efficient gas exchange structures.

Lungs are efficient gas exchange structures because:

1 **The surface area** for gas exchange is very large (due to millions of air sacs).

2 The gas exchange surface (walls of air sacs) is **very thin** allowing oxygen and carbon dioxide to pass through quickly by diffusion.

3 The gas exchange surface is **moist** so oxygen and carbon dioxide can dissolve so that diffusion can take place.

4 Millions of capillaries provide a good **blood supply** to the air sacs.

Gas exchange in the air sacs

C

C4 Describe gas exchange between the air sacs and the surrounding blood vessels.

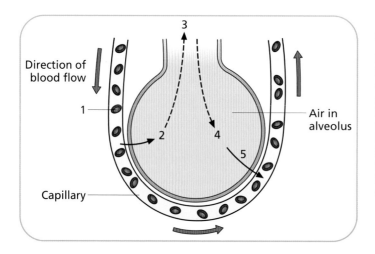

Figure 5.6 Gas exchange in the air sacs

Blood in the capillary at (1) is low in oxygen but high in carbon dioxide, so carbon dioxide diffuses out of the blood and into the air sac (2), then passes up the bronchioles, bronchus and trachea as we breathe out (3). Oxygen follows the opposite path and diffuses from the air sac (4) into the capillary (5).

Structure and function of the heart

G

G4 Identify the four chambers of the heart.

G

G5 Describe the path of blood flow through the heart and blood vessels connected to it.

G

G6 Describe the positions and functions of the heart valves.

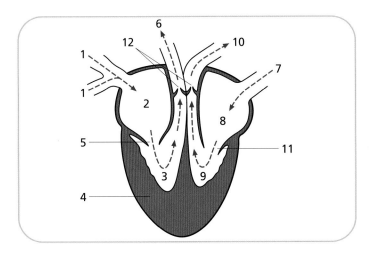

Figure 5.7 **Structure of the heart**

Blood takes the following path as it passes through the heart.

Blood returns from the head, arms and rest of the body through a large vein, the **vena cava** (1).

Blood enters the **right atrium** (2). When it is full, the muscular wall squeezes blood into the **right ventricle** (3). When the ventricle is full the thick muscular wall (4) contracts. The heart valve (5) closes and stops blood going back to the atrium.

Blood leaves the heart through the **pulmonary artery** (6) and goes to the lungs.

After passing through the lungs, blood returns to the heart through the **pulmonary vein** (7) and enters the **left atrium** (8). It squeezes blood into the **left ventricle** (9). When full, the ventricle pumps blood out of the heart through the **aorta** (10). Another valve (11) stops blood going back into the left atrium.

The pulmonary artery and aorta also have valves (12) to stop blood flowing back into the heart between beats.

Left and right are reversed in a diagram of the heart.

The four chambers of the heart are:

The right atrium (2), **the right ventricle** (3), **the left atrium** (8), **and the left ventricle** (9).

The heart – a living pump

G7 Explain the difference in thickness of the walls of the ventricles.

G8 State that the heart obtains its blood supply from coronary arteries.

G9 State that blood leaves the heart in arteries, flows through capillaries and returns to the heart in veins.

The heart is really two pumps in one. The right side pumps blood to the lungs and back, the left side pumps blood all round the rest of the body and back.

The muscles of the left ventricle are thicker than those of the right ventricle, as greater pressure is needed to pump blood all round the body.

Blood leaves the heart through **arteries**, passes through **capillaries** in the organs of the body, then returns to the heart through **veins**.

The heart is a muscular bag and needs a good supply of oxygen and food. This is provided by the **coronary arteries** which branch off from the aorta.

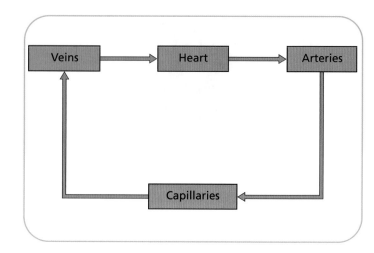

Figure 5.8 Movement of blood round the body

What causes the pulse?

G10 State that the pulse indicates that blood is flowing through an artery.

Each time the heart beats (about 70 times a minute on average), blood is pushed into the arteries under high pressure. This causes the walls of the arteries to bulge as the blood flows through them. Each heartbeat can be felt as a pulse in the arteries.

Blood

G11 Describe the function of red blood cells and plasma in the transport of respiratory gases and food.

Blood consists of **red cells** and **white cells** carried in a liquid called **plasma**.

Red Cells

Size – tiny (about 5.5 million per cubic millimetre).

Filled with red pigment **haemoglobin**.

Function – to transport oxygen from the lungs to the tissues.

Unusual in having no nucleus.

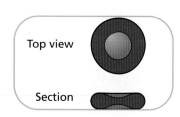

Figure 5.9 Features of red blood cells

Plasma

Yellowish coloured liquid.

Functions:

1   to transport cells

2   to carry dissolved food (in particular glucose and amino acids) from the intestines to the rest of body

3   to carry dissolved carbon dioxide from tissues to the lungs for removal from the body.

The function of haemoglobin

C5 Explain the function of haemoglobin in the transport of oxygen.

Haemoglobin is the red pigment in red blood cells responsible for transporting oxygen from the lungs to the tissues.

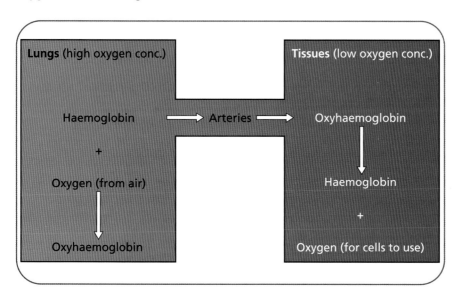

Figure 5.10 The function of haemoglobin

Gas exchange between blood and cells

G12 Describe gas exchange between the body cells and the surrounding capillaries.

Blood entering capillaries in the tissues is saturated with oxygen. The cells near the capillaries are low in oxygen because they use it up as soon as they get it. This causes a **concentration gradient** between blood and cells, so oxygen rapidly diffuses from the blood into the cells.

HOW TO PASS STANDARD GRADE BIOLOGY

C

Efficient gas exchange between blood and cells

C6 Describe the features of a capillary network which allow efficient gas exchange.

A **large surface area** due to arteries dividing many times into **tiny capillaries** and the **thin walls** (one cell thick) of capillaries, make them very efficient at exchanging oxygen and carbon dioxide between the blood and tissues.

G

Questions

1. Oxygen and carbon dioxide levels were measured in two air samples.

    Sample A – oxygen 16.5%; carbon dioxide 4.1%.

    Sample B – oxygen 20.4%; carbon dioxide 0.04%.

    Which sample was exhaled air?

2. (a) Rearrange the following parts to indicate the path of air during breathing in:
    bronchioles; trachea; air sacs; mouth and nose; bronchi.

    (b) In which of the parts listed in 2(a) does gas exchange take place?

3. Rearrange the following parts to indicate movement of blood through the heart, lungs and associated vessels (start with the vena cava):
    vena cava; left ventricle; lungs; right atrium; aorta; pulmonary artery; right ventricle; pulmonary vein; left atrium.

C

4. Copy the following sentences, choosing correct alternatives:

    During breathing in, intercostal muscles contract/relax and the rib cage is raised/lowered. During breathing out, intercostal muscles contract/relax and the rib cage is raised/lowered.

5. Describe three features of the lungs that make them efficient gas exchange structures.

6. Complete the following word equation:

    haemoglobin + oxygen →

Coordination

Two eyes are better than one

G1 State that judgement of distance is more accurate using two eyes rather than one.

Any activity that requires judgement of distance (e.g. catching a ball, driving etc.) is more accurate with two eyes than with only one.

Binocular vision

C1 Explain the relationship between judgement of distance and binocular vision.

Each eye forms a slightly different image due to being several centimetres apart. The brain therefore receives these two images and combines them into one. From the difference between the two images of an object, the brain is able to estimate how far away the object is. If the object is moving towards or away from us, binocular vision helps to estimate speed of movement.

Structure and function of the eye

G2 Identify the cornea, iris, lens, retina, optic nerve and state their functions.

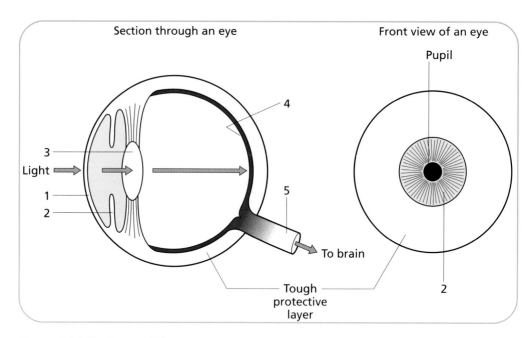

Figure 5.11 Features of the eye

Light enters the eye through the transparent **cornea** (1). The amount of light entering the eye is controlled by the **iris** (2). Light passes through the hole in the middle of the iris (the pupil) and is focussed by the **lens** (3). Light forms an image on a thin layer of light-sensitive cells at the back of the eye, the **retina** (4). Nerve impulses from these cells pass along the **optic nerve** (5) to the brain.

Hearing

G3 State that judgement of direction of sound is more accurate using two ears rather than one.

With sight, distance judgement is better with two eyes, and with hearing, judgement of direction of sound is better with two ears than it would be with only one.

Structure and function of the ear

G4 Identify the ear drum, middle ear bones, cochlea, auditory nerve and semi-circular canals and state their functions.

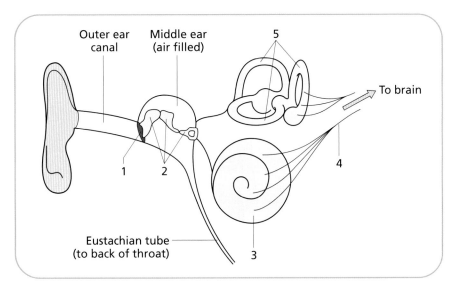

Figure 5.12 Features of the ear

Sound waves pass down the outer ear canal and hit the **ear drum** (1) causing it to vibrate. The vibrations are carried by the tiny **middle ear bones** (2) to the fluid-filled inner ear. Nerves in the coiled **cochlea** (3) detect the vibrations in the fluid and send impulses along the **auditory nerve** (4) to the brain.

The three **semi-circular canals** (5) detect movement not sound. When the head moves, nerve impulses are sent to the brain giving information about movement which helps us balance.

Semi-circular canals

C2 Explain how the arrangement of semi-circular canals is related to their function.

The three semi-circular canals are at right angles to each other, so whichever direction the head moves, at least one is stimulated. The three directions detected are:

1   up and down (nodding head)

2   rotating (shaking head)

3   side to side (ear down towards the shoulder).

The nervous system

(G) **G5** State that the nervous system is composed of the brain, spinal cord and nerves.

(G) **G6** State that nerves carry information from the senses to the central nervous system and from the central nervous system to the muscles.

The brain and spinal cord make up the **central nervous system** (CNS).

Sensory nerves relay information from the sense organs to the central nervous system. **Motor nerves** carry impulses from the CNS to the muscles causing them to contract.

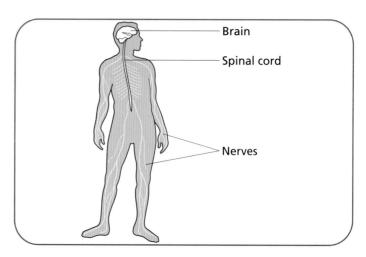

Figure 5.13 **The central nervous system**

The reflex arc

C **C3 Describe how a reflex action works, using a simple model of a reflex arc.**

Reflex actions are fast movements, usually to prevent damage occurring to the body. Nerve impulses travel from a sense organ to the spinal cord which causes a very fast muscle contraction.

Example

Jerking a hand away from a hot object.

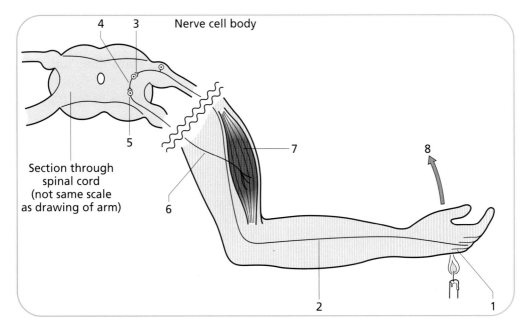

Figure 5.14 Features of a reflex arc

Pain receptors in the back of the hand are stimulated (1). A nerve impulse travels up a sensory neurone (2). The impulse crosses a synapse (junction) (3) to an intermediate neurone (4) inside the spinal cord. The impulse crosses another synapse (5) and travels down a motor neurone (6). The motor neurone stimulates a muscle in the arm to contract (7) which jerks the hand away from the flame (8). As the brain is not involved, this all happens very fast, reducing the damage done to the hand.

Responding to stimuli

C4 State that the CNS sorts out information from the senses and sends messages to those muscles which make the appropriate response.

The CNS receives information all the time (even when asleep) about our surroundings and about our internal environment. A lot of this information needs no action to be taken and is 'filtered out' before it reaches the conscious part of our brain (e.g. background noises that you only notice when they stop).

Some information must be acted upon very fast before the body is damaged (e.g. jerking a hand away from a hot object). This is called a reflex action and does not involve the brain.

Other information will be analysed by the brain before a conscious action is taken.

The brain

C5 Identify the cerebrum, cerebellum and medulla and state their function.

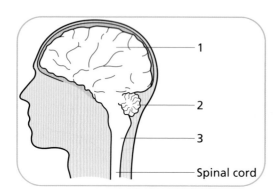

Spinal cord

Figure 5.15 Features of the brain

The **cerebrum** (1) is the part of the brain responsible for:

- thought;
- memory;
- interpreting information from the sense organs;
- movement of the body.

The **cerebellum** (2) is responsible for balance and fine control of muscles, making our movements precise and coordinated.

The **medulla** (3) looks after automatic functions of the body, such as controlling the heart beat, breathing etc.

Questions

1 Connect the part of the eye in list A with the correct function in list B.

A	B
Iris	Carries nerve impulses to the brain
Retina	Focuses light on the back of the eye
Optic nerve	Controls the amount of light entering the eye
Cornea	Light sensitive cells that form an image
Lens	Transparent protective layer

2 Rearrange the following parts of the ear to show the sequence of events from sound waves to the brain:

cochlea; middle ear bones; auditory nerve; ear drum

3 What is a reflex action?

4 Describe the sequence of events involved in the reflex arc responsible for pulling your hand away from a hot object.

5 Construct a table showing the functions of the cerebrum, the cerebellum and the medulla.

Changing Levels of Performance

Muscle fatigue

G1 State that continuous or rapidly repeated contraction of muscle results in fatigue.

G2 State that muscle fatigue results from a lack of oxygen and a build up of lactic acid.

If a muscle stays contracted for a long time, or contracts and relaxes rapidly for a long time, it begins to suffer from muscle fatigue (muscle tiredness).

Example

1 Continuous contraction: carrying a heavy shopping bag, your hands and arms get tired.
2 Rapidly repeated contraction: in a running race, your arms and leg muscles get tired.

Muscles need food and oxygen to work properly. In the examples above, these may not be supplied by the blood fast enough (particularly oxygen). If muscles contract when they are not getting enough oxygen, they work very inefficiently and lactic acid builds up in the muscle tissue. It is this substance that causes muscle fatigue.

Anaerobic respiration in muscles

C1 Explain muscle fatigue in terms of anaerobic respiration.

In aerobic respiration, energy is released from food with the help of oxygen. When muscles cannot get enough oxygen for aerobic respiration, they can use anaerobic respiration to release a little energy from food.

Disadvantages of anaerobic respiration

1 It is much less efficient than aerobic respiration (less energy is released from food).
2 It produces lactic acid which causes muscle fatigue.

Advantage of anaerobic respiration

It is able to keep cells alive for a short time in the absence of oxygen.

The effect of exercise on pulse rate and breathing rate

G G3 **Explain why pulse rate and breathing rate increase with exercise.**

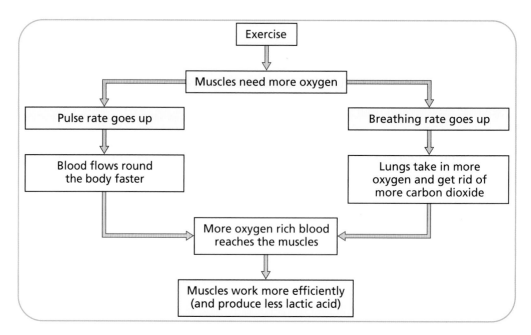

Figure 5.16 **Effects of exercise on pulse rate and breathing**

The effect of training

G G4 **State that with exercise, the pulse rate, breathing rate and lactic acid level rise less in an athlete than in an untrained person.**

An athlete is 'fitter' than an untrained person because heart rate and breathing rate rise less in an athlete and an athlete's muscles also produce less lactic acid.

Recovery time

G G5 **State that recovery time is the time taken to return to normal levels of pulse rate, breathing rate and lactic acid.**

G G6 **Describe how recovery time can be used as an indication of physical fitness.**

Recovery time is the time it takes, after exercise, for the pulse rate, breathing rate and lactic acid level to return to normal.

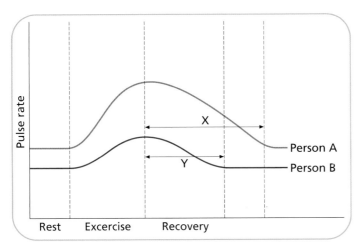

Figure 5.17 Graph showing recovery time from exercise

A fit person's recovery time will be shorter than that of an unfit person.
Notice that:

1   Person A has a higher resting pulse rate (unreliable indicator).
2   A's pulse rate rises more than B's during the period of exercise.
3   A's recovery time (X) is longer than B's (Y). This indicates that B is fitter
    than A.

1   Recovery time can also apply to breathing rate and lactic acid level.
2   This comparison can only be made if the two people are similar in size, age, sex
    and do identical exercises for the same time.

Improving recovery time

C2 State that training improves the efficiency of the lungs and circulation.

**C3 Explain the relationship between the effects of training and recovery
time.**

Training improves the efficiency of the body in the following ways:

1   The lungs absorb oxygen and pass it to the blood faster (removing carbon
    dioxide from the blood faster also).
2   The heart increases in size and strength and so pumps blood round the body
    faster.

These two effects mean that after exercise, oxygen can be quickly supplied to
muscles to break down the lactic acid that has built up during anaerobic conditions.
This is called **paying back the 'oxygen debt'**. The faster this is done, the shorter
is the recovery time.

HOW TO PASS STANDARD GRADE BIOLOGY

G

C

Questions

1 Explain how continuous muscle contraction leads to muscle fatigue.

2 Explain how increased pulse rate and breathing rate can reduce muscle fatigue during exercise.

3 What is meant by recovery time?

4 Explain how an athlete, by training, can improve his recovery time.

Chapter 6

INHERITANCE

> ### Variation

What is a 'species'?

G1 **State that a species is a group of interbreeding organisms whose offspring are fertile.**

Similar animals (or plants) may breed with each other. If they produce offspring that are fertile (i.e. can also reproduce) they are said to belong to the same **species**.

Organisms that cannot breed with each other, or do breed, but produce infertile offspring* are said to be members of **different species** [*e.g. horse × donkey → mule (sterile offspring)].

Variation within species

G2 **State that variation can occur within a species.**

Although members of a species show many similarities (characteristics of the species), individuals will show many differences (e.g. tough spiky leaves are characteristics of holly, but the number of spikes and colour of the leaf may vary).

Types of variation

G3 **Give examples of continuous and discontinuous variation.**

Continuous variation:

A feature of an animal or plant such as length or weight, that can be anywhere in a range from the smallest to the largest.

Example

1　Height of sweet pea plants.
2　Weights of people in a slimming club.
3　Heights of people on a bus.
4　Length of mouse tails.

Discontinuous variation:
A feature of an animal or plant that falls into two or more distinct groups.

Example

1 Flower colour of sweet peas.

2 Attached or unattached ear lobes in humans.

3 Human blood groups (A, B, AB or 0).

4 Smooth or wrinkled maize kernels.

Continuous or discontinuous?

C

C1 Explain what is meant by continuous and discontinuous variation.

Continuous variation

Example

Height of sweet pea plants.

In a group of sweet pea plants, one will be the smallest and one will be the tallest. In between, the other plants will form a continuous range of sizes.

This is called a normal distribution curve. It means that a few will be small plants, a few will be tall plants but the commonest height will be somewhere in between. Continuous variation is often shown by a **histogram** which is a bar graph with increasing values along the X axis.

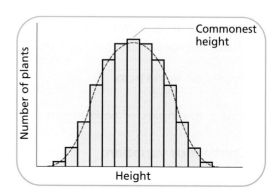

Figure 6.1 Histogram showing continuous variation

Discontinuous variation
Every member of the population must belong to one or other of the possible groups.

Example

Flower colour of sweet peas.

Discontinuous variation is shown by a **bar graph**.

When trying to decide if an example is continuous or discontinuous, ask yourself these two questions:

1. Can I separate the example into a number of separate categories? If the answer is yes, it is discontinuous. If the answer is no, it is continuous.

2. Would the example best be shown as a histogram? If the answer is yes, it is continuous. If it would be best as a bar graph (for example if there are no increasing values, but simply different categories) it is discontinuous.

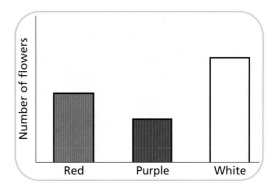

Figure 6.2 Bar graph showing discontinuous variation

Questions

1. Explain why horses and donkeys are said to belong to different species.

2. Arrange the following into examples of continuous and discontinuous variation:
weight of acorns; length of pea pods; colours of pea flowers; times for athletes to run 100 m.

What is Inheritance?

Inherited characteristics

G **G1** State that certain characteristics are determined by genetic information received from the parents and give examples from animals and plants.

Each of us has characteristics that can be seen in our parents and grandparents. This is also true of other animals and plants. This is because 'information' which determines these characteristics is passed on from parents to offspring. The nature of this genetic information is explained later in this chapter.

Examples of inherited characteristics:

Example

1   Humans: tongue rolling, red hair, shape of nose etc.

2   Animals: fur colour, eye colour in fruit flies etc.

3   Plants: leaf shape, flower colour in peas, tallness/dwarfness in peas etc.

Different phenotypes

G2 Identify examples of phenotypes of the same characteristic.

Most physical characteristics can appear in a number of different forms. These different forms are called **phenotypes**. See the table below for examples.

Table 6.1 **Phenotypes**

	Characteristic	Different phenotypes
Humans	Hair colour Hair type	Fair, dark, red Straight, wavy, curly
Animals	Wing shape in fruit flies Fur colour in mice	Straight, curved Black, white, grey
Plants	Height of pea plants Colour of maize kernels	Tall, dwarf Yellow, purple

True-breeding characteristics

G3 Identify examples of true-breeding, dominant and recessive characteristics from the numbers and phenotypes of given crosses.

G4 Identify generations as P, F1 and F2 from given examples of crosses.

Pollen from the flowers of a tall pea plant is used to pollinate flowers on another tall pea plant. The seeds are later collected and planted to produce the next generation. This can be shown as follows:

Parents (P) Tall pea plant × Tall pea plant

↓

First new generation (F1)   All tall

'F' stands for filial or 'daughter' generation

If two of these F1 plants are crossed:

Tall pea plant × Tall pea plant

↓

Second new generation (F2)   Again all tall

When a characteristic is passed on unchanged like this, generation after generation, the plants (or animals) are said to be **true-breeding** for that characteristic.

But if a tall pea plant is crossed with a dwarf pea plant:

P   Tall pea plant × dwarf pea plant

↓

F1   All tall just as before

This shows that when the two phenotypes are 'mixed' in the F1 generation, only the tall phenotype seems to operate. This is called the **dominant** phenotype. Dwarf does not operate when mixed with tall and so is called the **recessive** phenotype.

If two of the F1 plants are now crossed:

Tall pea plant × Tall pea plant

↓

F2   three tall plants to every one dwarf plant (i.e. tall to dwarf in a ratio of 3:1)

HOW TO PASS STANDARD GRADE BIOLOGY

Pea plants will probably produce more than four offspring, but however many they produce there should be roughly three times more tall plants than dwarf.

This shows that the recessive phenotype can appear again in the F2 generation.

True-breeding crosses

C1 State that the parents in experimental monohybrid crosses are usually true-breeding and show different phenotypes of the same characteristic.

C2 Predict the proportions of the phenotypes of the F2 offspring of a monohybrid cross.

C3 Explain monohybrid crosses in terms of genotypes.

The original parents in a monohybrid cross are usually true-breeding. From a given cross, you should be able to work out the genotype of parents (P), gametes, F1 offspring and by crossing two of the F1, the genotypes, phenotypes and ratios of the F2 offspring.

e.g. Tall pea × dwarf pea (tall is dominant).

Capital letters are used to indicate the dominant phenotype (e.g. T. Small t is used for the recessive in this case).

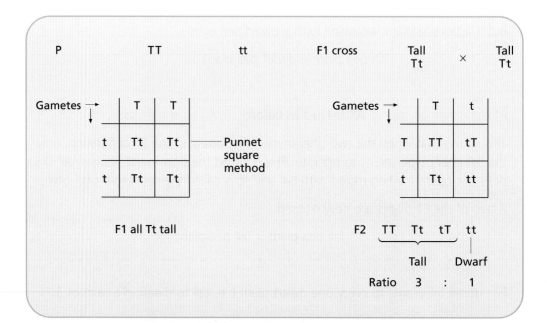

Figure 6.3 Predicting the F2 offspring of a monohybrid cross

Summary – so far

G5 State that the phenotypes of the F1 in a true-breeding cross are uniform.

In a true-breeding cross, all the offspring in the F1 and F2 generations will be the same as the parents.

In a non-true-breeding cross, all the F1 offspring will show the dominant phenotype, but in the F2 generation there will be a 3:1 ratio of dominant to recessive.

Chromosome number

G6 State that each body cell has two matching sets of chromosomes.

The nucleus of a human body cell has 46 chromosomes. This is made up of two matching sets of 23 chromosomes. Other organisms have different numbers, but they always consist of two matching sets.

G7 State that sex cells are called gametes.

G8 State that the reduction of the number of chromosomes to a single set occurs during gamete formation.

G9 State that each sex cell carries one set of chromosomes.

The sex cells, or gametes, involved in sexual reproduction are formed by a special kind of cell division. This separates the two sets of chromosomes so that one complete set goes into each gamete.

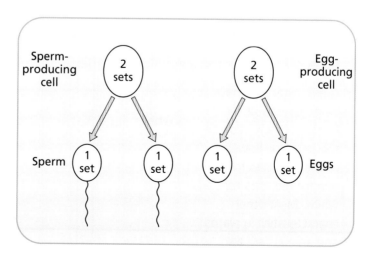

Figure 6.4 **Chromosome sets in sex cells**

Completing the double set again

G10 Describe how a complete double set of chromosomes is achieved at fertilisation.

Fertilisation occurs when a male gamete joins with a female gamete. Each has a single set of chromosomes, so the fertilised egg cell then has two sets once again, one set from each parent.

Exactly the same happens during fertilisation in plants.

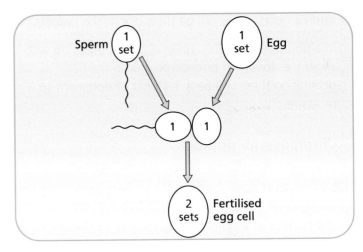

Figure 6.5 **Chromosome sets after fertilisation**

The role of genes

G11 State that genes are parts of chromosomes.

G12 State that a characteristic is controlled by two forms of a gene.

Information passed from parent to offspring is carried on the chromosomes. Each characteristic (e.g. hair colour, wing shape, flower colour etc.) is controlled by a section of a chromosome called a gene.

A pea plant may have white flowers or red flowers. This is because the gene for flower colour is in two forms: one for red, one for white. Most other characteristics in plants and animals are also controlled by two forms of a gene.

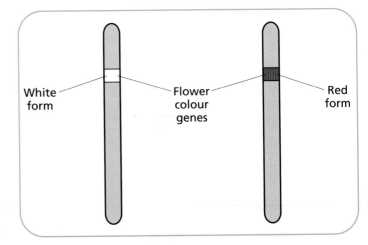

Figure 6.6 **Chromosomes from a pea plant**

Alleles

C4 State that different forms of a gene are called alleles.

Whenever a gene exists in more than one form, these different forms are called alleles (e.g. long and short are two alleles of the fruit fly gene for wing length, for other examples see the table 6.2 below).

Table 6.2 **Alleles**

Characteristic	Alleles
Pea flower colour	Red, white
Pea height	Tall, dwarf
Human colour vision	Normal, colourblind

The monohybrid cross

G13 State that each parent contributes one of the two forms.

G14 State that each gamete carries one of the two forms of the gene.

G15 State the meaning of the word genotype.

An experimental cross in which the inheritance of a single characteristic is being investigated is called a **monohybrid cross**.

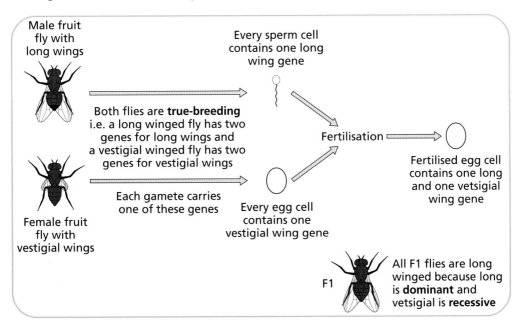

Figure 6.7 Monohybrid cross in fruit flies

If we now cross two of the F1 flies:

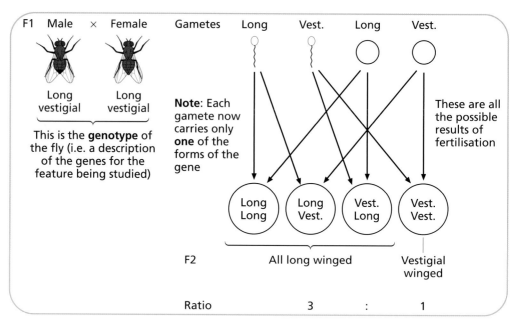

Figure 6.8 **Fly crosses**

In each generation, an individual will inherit one form of every gene from its 'father' and the other from its 'mother'.

Differences between observed and predicted figures

C5 Explain differences between observed and predicted figures in monohybrid crosses.

The F2 generation of the pea cross in C3 predicts a ratio of tall to dwarf of 3:1.

In an actual cross with many plants, the results were 265 tall plants and 82 dwarf plants. This is a ratio of 3.2:1

Actual ratios may differ from predicted ones for the following reasons:

1   The random nature of fertilisation.

2   Failure of seeds to germinate (in plants).

3   Death of seedlings/embryos.

Determination of sex

G16 State that the sex of a child is determined by specific chromosomes called X and Y chromosomes.

G17 State that in humans, each male gamete may have an X or a Y chromosome, while each female gamete has an X chromosome.

One of the 23 pairs of human chromosomes determines the sex of a person. A female has two identical chromosomes, called the X chromosomes (i.e. the female genotype is XX).

A male has two different chromosomes. One an X, the other a Y (i.e. the male genotype is XY).

G18 Explain how the sex of a child is determined with reference to the X and Y chromosomes.

The female gametes will all contain an X chromosome. Half of the male gametes will contain an X and half will contain a Y. Thus it is the father's gamete which determines the sex of a child.

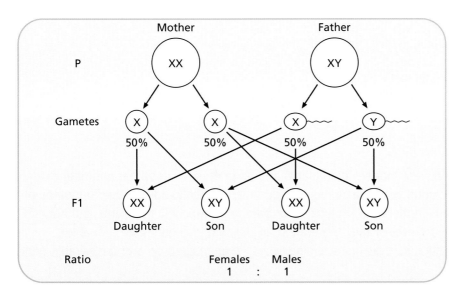

Figure 6.9 Determination of sex

HOW TO PASS STANDARD GRADE BIOLOGY

G

C

Questions

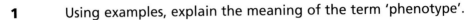

1   Using examples, explain the meaning of the term 'phenotype'.

2   Explain what is meant by an animal being 'true-breeding' for a certain characteristic.

3   How many chromosomes are there in (a) a human body cell and (b) a human gamete?

4   Why is it essential that gametes carry a half set of chromosomes rather than a full set?

5   Explain the meaning of the term 'genotype'.

6   Explain why it is said that 'the father determines the sex of a child'.

7   Calculate the genotypes and phenotypes resulting from the following cross:
a long-winged fruit fly (Ll) × a vestigial-winged fly (ll)

8   Explain the meaning of the term 'allele' using an example.

9   Give two reasons why actual (observed) results of a cross may differ from predicted results.

⇨   *Genetics and Society*

Selective breeding

G   **G1 Give two examples of an improved characteristic resulting from selective breeding.**

The ancestors of modern crop plants and animals (such as wheat and beef cattle) were much poorer in the quality and quantity of their produce. Over hundreds of generations farmers have selected, for breeding purposes, the plants and animals that were the best of their generation. In this way, the genes for desirable characteristics, such as high-yielding wheat, were passed on and undesirable genes were lost.

Other examples of characteristics improved by selective breeding include:

Example

1   Increased disease resistance in crop plants such as wheat.
2   Increased milk yield in dairy cattle.
3   Increased meat production in beef cattle.
4   Leaner meat (i.e. less fat) produced by beef cattle.

Enhancing characteristics by selective breeding

C1 Describe two examples, one plant, one animal, of the enhancement of a characteristic through selective breeding.

Plants

Four popular vegetables, cabbage, cauliflower, broccoli and brussel sprouts, have all been produced by selective breeding from a common ancestor. This has taken many generations.

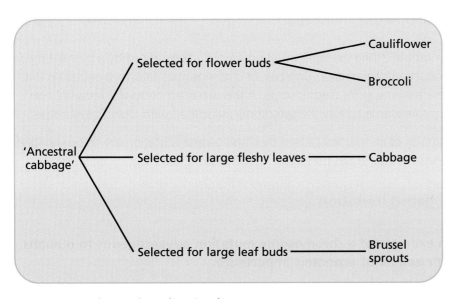

Figure 6.10 Selective breeding in plants

Animals

Different varieties of cattle have been produced through hundreds of years of selective breeding.

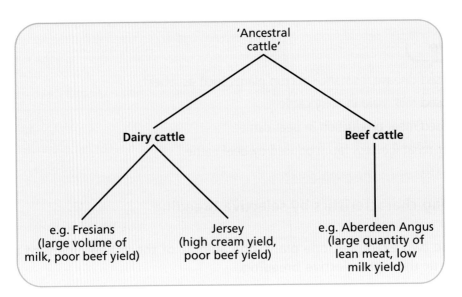

Figure 6.11 Selective breeding in animals

Chromosome mutation in humans

G **G2 Describe an example of a human condition caused by a chromosome mutation e.g. Down's syndrome.**

Down's syndrome is one example of a human condition caused by a mistake made during production of gametes. One pair of chromosomes fails to separate so that one gamete has one extra chromosome. If this is passed on to the fertilised egg, the extra chromosome causes the symptoms associated with Down's syndrome.

Other examples of conditions caused by chromosome mutation are colour blindness and Albinism (an albino has no pigmentation in skin, hair etc.).

Advantageous mutation

C **C2 Give an example of a chromosome mutation advantageous to humans, in a plant or animal of economic importance.**

Most mutations are harmful and the individual may not survive. Sometimes mutations can be beneficial and so the new characteristic is passed on to future generations.

There are examples of beneficial mutations that have economic importance. A bacterium that can digest oil can help to clear up oil spills. Modern bread wheat has multiple sets of chromosomes due to mutation. This gives it many advantages over wild wheat, such as higher yield.

Rate of mutation

C3 Give an example of a factor which can influence the rate of mutation in an organism.

Chromosomes may mutate spontaneously (i.e. without any external cause). Most mutations, however, are due to some environmental factor. Such a factor is called a mutagen and it may work by changing the structure of a chromosome, or by damaging the mechanism which separates the sets of chromosomes during gamete formation. Down's syndrome is an example of the latter.

Radiation, e.g. X-rays, ultra-violet, radiation from a radioactive material, and chemicals, e.g. mustard gas, benzene, are examples of mutagens.

Amniocentesis

G3 State that amniocentesis can be used to detect chromosome characteristics before birth.

If it is thought that a human embryo may have a genetic problem, such as Down's syndrome, the chromosome characteristics of the embryo can be checked well before birth.

An amniocentesis is carried out by taking a sample of the amniotic fluid surrounding the embryo and looking at the embryo's cells that are in the fluid. The nucleus of a cell is photographed through a microscope and the chromosomes can be checked. If a serious abnormality is found, the mother may decide to have an abortion.

Questions

1   Give one example for plants and one for animals, of an improvement made by generations of selective breeding.

2   Explain what is meant by a 'mutagen' and give an example.

Chapter 7

BIOTECHNOLOGY

Living Factories

Biotechnology is – using living organisms to:

1   produce a product for human consumption or

2   carry out a process, of benefit to humans.

The first part of this chapter deals mostly with fermentation by yeast – a form of biotechnology that has been known for thousands of years.

'Traditional' biotechnology

G **G1 State that the raising of dough and the manufacture of beer and wine depend on the activities of yeast.**

G **G2 Identify yeast as a single-celled fungus, which can use sugar as food.**

Yeast is a single-celled fungus used in bread making and also in making beer, wine and other alcoholic drinks.

Yeast uses sugar for food and it produces carbon dioxide gas (which makes dough rise) and alcohol as waste products.

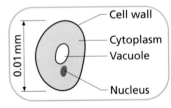

Figure 7.1 Features of a yeast cell

Anaerobic respiration

C **C1 Describe the process of anaerobic respiration and compare it with aerobic respiration.**

Respiration (see p. 85) is the release of energy from food and takes place in every living cell. Most cells use oxygen to make this energy release efficient. This is **aerobic respiration.**

Some cells, however, can respire without oxygen. This is called **anaerobic respiration** (AN = non) and it is less efficient as it releases less energy from the food than aerobic respiration. Fermentation by yeast is an example of anaerobic respiration.

Word equations

1   Aerobic respiration

glucose + oxygen → carbon dioxide + water + energy

2   Anaerobic Respiration (in plants and yeast)

glucose → carbon dioxide + ethanol + energy

One proof that anaerobic respiration is less efficient than aerobic is that ethanol produced in anaerobic respiration still contains a lot of useful energy, whereas water produced in aerobic respiration contains none.

Commercial brewing

C2 Describe how commercial brewers provide the best conditions for yeast.

For efficient brewing, the best conditions for yeast must be provided. These are:

1   plenty of food
2   the correct temperature
3   sterile conditions.

Food

This is sugar, usually in the form of maltose, produced by germinating barley.

Temperature

The fermentation process gives off heat, so the brewer must cool the fermentation vessel to the correct temperature.

Sterile conditions

Any wild yeast or bacteria getting into the brew would spoil it, so all parts of the brewing process must be kept sterile, usually by steam cleaning before the brew is started.

Batch processing

C3 Explain what is meant by the term 'batch processing'.

Traditional biotechnology such as brewing uses batch processing in which all the raw materials are put into the fermentation vessel and later the product is removed.

One disadvantage of this system is that it must be thoroughly cleaned out between batches. This is time in which no product is being made and so no money is being earned.

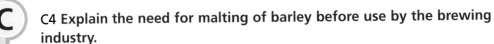

Malting barley

C **C4 Explain the need for malting of barley before use by the brewing industry.**

Barley is used as the source of food for alcoholic fermentation by yeast in the brewing industry (and as the first stage of whisky making). The carbohydrate in barley is in the form of starch which must be converted to sugar before yeast can use it.

Enzymes in the barley grains convert the starch to maltose (or 'malt') during germination. In the brewing industry this process is called **malting** the barley.

Fermentation by yeast

G **G3 Using a word equation, state the process of fermentation of glucose by yeast.**

Alcoholic fermentation is the name given to the process in which yeast uses sugar and produces carbon dioxide and alcohol.

glucose (sugar) $\rightarrow$ carbon dioxide + ethanol (alcohol) + energy

Cheese and yoghurt

G **G4 State that the manufacture of cheese and yoghurt depends on the activities of bacteria.**

Another traditional type of biotechnology is the use of bacteria in making cheese and yoghurt.

Cheese

In cheese making, bacteria are used to:

1   produce a chemical to clot milk
2   ripen the cheese
3   flavour the cheese.

Blue cheeses (e.g. Stilton) have fungi added to them to produce the blue veins and characteristic flavour.

Yoghurt

Bacteria are used to:

1   clot or thicken the milk
2   produce lactic acid which gives the 'sharp' flavour of yoghurt.

G5 State that the souring of milk is a fermentation process.

The production of lactic acid by milk bacteria is another type of fermentation.

Lactic fermentation

C5 Explain the souring of milk in terms of bacterial fermentation of lactose.

A form of anaerobic fermentation by milk bacteria turns lactose (milk sugar) into lactic acid. This gives sour milk its acid taste and causes coagulation of the milk proteins.

Questions

1. Alcoholic fermentation is an 'anaerobic process'. What does this term mean?

2. Write a word equation to describe alcoholic fermentation by yeast.

3. Describe the uses of bacteria in cheese making.

4. Describe the differences between alcoholic fermentation in yeast and anaerobic respiration in animal cells.

5. Describe what happens during 'malting' of barley and explain why it is necessary before barley is used in the brewing industry.

Problems and Profit with Waste

The environmental effect of sewage

G1 Describe some examples of the damage done to the environment by disposal of untreated sewage.

G2 Give some examples of diseases which may be spread by untreated sewage.

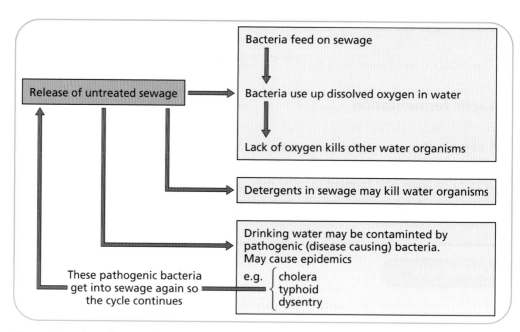

Figure 7.2 The effects of sewage on the environment

Handling micro-organisms in the laboratory

G G3 Describe the precautions that must be taken during laboratory work with micro-organisms.

G G4 Explain the importance of such precautions in any biotechnological work.

Table 7.1 Necessary precautions in the laboratory

Precautions	Reasons for precautions
Hands washed before and after work	Remove bacteria picked up from the environment/remove any picked up during lab work
Benches swabbed with disinfectant	Remove bacteria and spores from benches
Lab coat should be worn	Protect clothes from bacteria
Use only safe bacterial cultures	Other sources, e.g. air, soil, etc. may contain pathogenic bacteria
Autoclave all equipment	Heat in an autoclave kills all foreign bacteria
Work beside Bunsen, flaming loops and necks of culture bottles.	Prevent entry of foreign bacteria and kill cultured bacteria on loops etc.
Incubate bacterial cultures below body temperature	This discourages the growth of pathogens which grow best at 37°C
Autoclave all equipment after use and cultures before disposal	Kill all bacteria in case any pathogens have appeared

Handling micro-organisms in industry

C1 Explain the precautions taken during manufacturing processes with reference to resistant fungal and bacterial spores.

In most industrial processes involving micro-organisms, the greatest danger is of foreign bacteria or fungi getting into the process rather than the cultured micro-organisms escaping.

Some bacteria and fungi make spores (such as seeds) that are resistant to drying out and to heat. If they get into the manufacturing process they may cause:

1   a health hazard – they may be pathogenic

2   financial loss if a whole batch of product is contaminated.

The usual precaution taken to prevent this from happening is for all equipment (fermenters, pipework etc.) to be frequently steam-cleaned. The high temperature of the steam can kill the spores.

Sewage treatment – bacterial decay in action

G5 State that sewage treatment includes its breakdown by decay micro-organisms to products harmless to the environment.

One of the main dangers of releasing untreated sewage is that it provides ideal food for bacterial growth. This may then harm the environment (see G1 and G2, pp. 121–2).

If this bacterial growth is contained within a sewage treatment plant, the environmental damage is avoided.

Aerobic breakdown of sewage

C2 Explain why complete breakdown of sewage is only possible in aerobic conditions.

Anaerobic respiration (without oxygen) is much less efficient than aerobic respiration (with oxygen). Waste products of anaerobic respiration still contain a lot of energy. In other words the breakdown of the food is not complete.

Breakdown of sewage by anaerobic bacteria will always leave some sewage material untreated.

Aerobic bacteria, however, will complete the breakdown, leaving only carbon dioxide and water.

Sewage with **anaerobic** bacteria → partially digested sewage (unsafe)

Sewage + oxygen with **aerobic** bacteria → carbon dioxide + water (safe)

G

G6 Describe how the oxygen required by micro-organisms can be provided during sewage treatment.

'Sewage-eating' bacteria need oxygen. This can be provided in three ways:

1   Trickling the sewage through filter beds. Air spaces between the stones provide oxygen.
2   Bubbling compressed air through tanks containing activated sludge (sewage with added sewage-eating bacteria).
3   Agitation of activated sludge. Stirring mixes in oxygen from the air.

C

C3 Explain why a range of micro-organisms is needed to break down the range of materials in sewage.

Sewage consists of many different types of waste material (carbohydrates, proteins, fats and minerals).

Each type of micro-organism tends to digest just one type.

Many different types of micro-organism are therefore needed to complete the treatment.

Making use of waste

G

G7 Give two examples of useful products made from waste by micro-organisms. Explain the economic importance of this technology.

Industry can save money by using micro-organisms to clean up waste that would otherwise need expensive treatment.

Industry can make money by selling the product made by some of these micro-organisms.

Example

1   Animal food can be made from yeast grown on waste sugar solution.
2   Biogas (methane) can be made by decay of bacteria in domestic waste dumps and on farms, from animal waste.

BIOTECHNOLOGY

Advantages of upgrading waste

C4 Explain the advantages of upgrading waste in terms of increasing its available energy or protein levels.

Micro-organisms can take low-grade waste with little energy or nutrient value and convert it to high-grade material with a higher energy content or nutrient value.

e.g. weak sugar solution (low energy content, low nutrient value)

↓

used as food by yeast

↓

yeast harvested (high protein content makes it a good animal food)

Fuel from micro-organisms

G8 State that alcohol and methane are products of fermentation.

G9 Explain the advantages of fermented fuel compared with fossil fuel.

A fuel is something that can be burned to release energy. Alcohol and methane are two fuels produced by a fermentation process carried out by micro-organisms in the absence of oxygen.

Alcohol – produced by fermentation of sugar by yeast.

Methane – produced by fermentation of waste by bacteria.

Fermented fuel has several advantages over fossil fuel (coal, oil and gas):

1   Fermented fuel is a renewable supply whereas fossil fuel is non-renewable (i.e. one day they will run out).

2   Using fermented fuel for energy helps to preserve oil etc. for use as raw materials in the chemical industry.

The growth and harvesting of micro-organisms

G10 State that under suitable conditions, micro-organisms can reproduce very rapidly by asexual means.

HOW TO PASS STANDARD GRADE BIOLOGY

G **G11** State that micro-organisms may be harvested to provide protein-rich food for animals or humans.

Given suitable conditions (warmth, food, moisture and oxygen) micro-organisms can reproduce very rapidly. This is an example of asexual reproduction.

In this way, a single bacterium can produce over 1 million bacteria after only 20 divisions. Under ideal conditions, bacteria can reproduce approximately every 20 to 30 minutes. At this rate, 1 million bacteria can be produced in under 10 hours.

Such rapid growth on relatively cheap and simple food, makes micro-organisms very suitable for the production of protein-rich foods for animals and humans.

Example)

1   Yeast for cattle food.
2   'Pruteen', a single-celled fungus for cattle food.

Bacteria and nutrient cycles

C **C5** Describe the part played by bacteria in the process of decay and recycling of carbon and nitrogen.

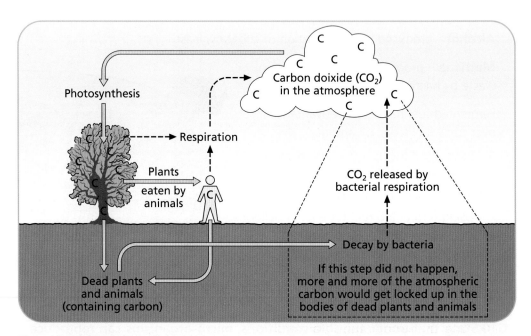

Figure 7.3 The carbon cycle

The Nitrogen cycle

The following diagram is reproduced from Chapter 1, p. 14. Here the emphasis is on the role of bacteria in the soil. Additional notes are provided below the diagram, relating to the processes numbered 1–4.

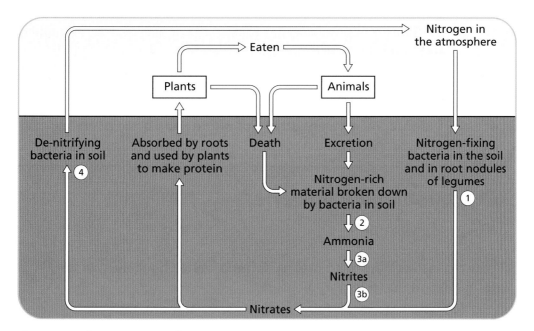

Figure 7.4 The nitrogen cycle

1   **Nitrogen-fixing bacteria**: found in soil and root nodules of legumes (i.e. peas, beans, clover etc.). They convert atmospheric nitrogen into **nitrates** which plants need to make protein.

2   **Decaying bacteria in soil**: they break down dead plants and animals and animal excretions into **ammonia**.

3   **Nitrifying bacteria**: there are two types:

(a) **nitrite bacteria** which convert ammonium compounds into **nitrites**.

(b) **nitrate bacteria** which convert nitrites into **nitrates** for use by plants.

4   **Denitifying bacteria**: they break down some of the useful nitrogen compounds in the soil and turn them back into atmospheric nitrogen.

Decay by micro-organisms

C6 Explain how micro-organisms break down material to provide energy.

All living organisms need energy to survive. Micro-organisms such as bacteria and fungi use living or dead plant and animal material as an energy source.

Questions

1 Give two examples of the conversion, by micro-organisms, of waste material into useful products.

2 Explain why the reproduction rate of micro-organisms makes them useful for the production of protein-rich foods. Give an example of such a food produced by micro-organisms.

3 How are unwanted micro-organisms prevented from contaminating biotechnological manufacturing processes?

4 Describe the different actions of nitrite and nitrate bacteria in the nitrogen cycle.

Reprogramming Microbes

Genetic engineering

G1 State that the normal control of bacterial activity depends on the information stored in its chromosomes.

Scientists have discovered ways of changing bacteria so that they produce substances needed by humans. This is done by altering the genes on the chromosomes which control the activity of the bacteria. This technique is called **genetic engineering**.

Transferring genes

G2 State that pieces of chromosome can be transferred from a different organism and so allow bacteria to make new substances.

Pieces of human chromosome can be transferred into a bacterial cell. The bacterium then makes the substance (e.g. insulin) made by the human genes.

Manipulating chromosomes

C1 Explain genetic engineering in terms of manipulation of chromosomal material.

Genetic engineering can alter an organism's chromosomes by:

1 adding new genes (often from a different organism)

2 removing undesirable genes (e.g. those that may cause a disease)

3 increasing the number of copies of a desirable gene already present in an organism.

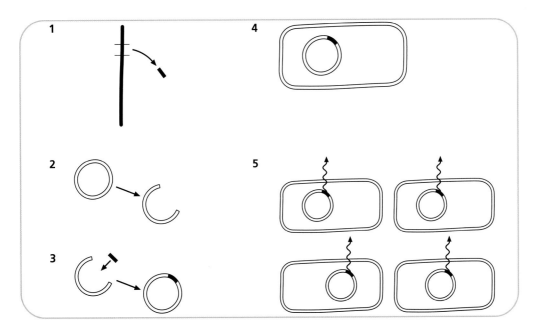

Figure 7.5 Transfering genes

1 Required genes cut from human chromosome by restriction enzymes.
2 Bacterial ring-like 'chromosome' (a plasmid) is cut open by enzymes.
3 The piece of human chromosome is inserted into the bacterial plasmid.
4 The plasmid is put back into a bacterial cell.
5 When the bacterium divides, each new cell contains the human genes and makes the desired substance which is released and can be collected.

C2 State that as a result of genetic engineering, bacteria may produce increased quantities of products and speed up processes.

If several copies of a desirable gene are made and put into a bacterial cell, even greater quantities of the desired substance will be produced. Sometimes a bacterium's own genes may be copied and increased to speed up a process carried out by the bacterium. For example, some bacteria can break down oil spills. If several copies of the gene involved are inserted, the bacterium will be able to clean up an oil spill much faster.

Genetic engineering compared with selective breeding

C3 Explain how genetic engineering can be better than selective breeding for the production of new organisms for a particular function.

If a new organism is needed to carry out a particular function, genetic engineering has the following advantages over selective breeding:

1 Genetic engineering is much faster. The new organism can be produced in just one or two generations. Selective breeding may take many generations.

i

2   Genetic engineering can give the new organism desirable characteristics that would never be developed by selective breeding (e.g. by incorporating genes from a different organism).

Genetic engineers must be extremely cautious when producing new strains of an organism, particularly bacteria, as they may prove to have unpredictable and damaging effects if they are ever released into the environment.

Products of genetic engineering

G

G3 Give some examples of the products of genetic engineering and their applications, e.g. insulin.

Table 7.2 Products of genetic engineering

Product	Application
Insulin	For human diabetics, to control blood sugar
Growth hormone	To encourage growth in abnormally small children
'Built-in' pesticides	Genes inserted into plant cells so that the plant produces its own insect-killing chemicals

The demand for insulin

C

C4 Explain the ever increasing need for insulin produced by biotechnology.

The hormone insulin is produced by the pancreas and controls the level of blood sugar. A diabetic is a person whose pancreas does not produce enough insulin. The number of diabetics is constantly rising due to people living longer and the human population growing.

Diabetics previously relied on animal insulin which was in limited supply and could cause side-effects. Human insulin can now be produced by genetically engineered bacteria.

'Biological' detergents

G

G4 State that 'biological' detergents contain enzymes produced by bacteria.

'Biological' detergents contain enzymes produced by genetically engineered bacteria. These washing powders are good at removing stains such as grass, blood etc. which contain biological material.

C5 Describe the advantages of using the low-temperature enzyme reactions of 'biological' detergents.

C6 Explain the action of 'biological' detergents in terms of digestion by enzymes.

'Biological' detergents have two other advantages over non-biological detergents. By working at low temperatures, they:

1   save energy – less money spent on heating water
2   do less damage to delicate fabrics that might be harmed in a hot wash.

The enzymes in biological detergents work by 'digesting' the biological material in the stain, making it easier to wash out. These detergents should be used at low temperatures, otherwise the enzymes may be destroyed.

Antibiotics

G5 State that an antibiotic is a chemical which prevents growth of micro-organisms.

An antibiotic is a chemical which prevents growth of micro-organisms. The term antibiotic is normally used to describe a drug which kills or prevents the growth of bacteria inside the human body.

C7 Explain why a range of antibiotics is needed in the treatment of bacterial diseases.

There are many different bacterial diseases which can infect different parts of the human body.

A particular antibiotic may work well against one type of bacteria but not against another, which is said to be resistant to the antibiotic. Therefore a range of antibiotics is needed to fight bacterial diseases.

Immobilizing enzymes

C8 Describe the advantages of immobilization techniques.

Enzymes are used in many manufacturing processes to convert raw materials into a product. Even with genetic engineering, enzymes may still be expensive to produce. To increase profit, the enzymes should be used many times if possible.

If the enzymes are just mixed with the substrate, it is difficult (and costly) to separate them from the end product and many enzymes will be lost.

Immobilization of the enzymes overcomes this problem.

One way of immobilizing enzymes is to attach the enzymes to glass beads so that they can be easily recovered and separated from the end product.

Continuous-flow processing by immobilized enzymes

C9 Explain continuous-flow processing by immobilized enzymes and the advantages this has over batch processing.

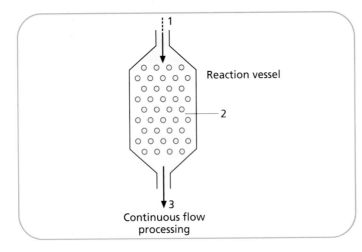

Continuous flow processing

Figure 7.6 Features of continuous-flow processing

Substrate (raw material) is poured in to the top of the reaction vessel in a steady stream (1).

Substrate trickles over enzymes held immobilized in the reaction vessel (2). The enzymes act on the substrate to produce the product which constantly trickles out at the bottom of the vessel (3).

Advantages over batch processing

1   It can be run for long periods without having to stop for cleaning (i.e. less time lost during which no product is being made).

2   The product does not have to be separated from the substrate or enzymes as may be necessary in batch processing.

Questions

1   Give two examples of products made by inserting genes from other organisms into bacteria.

2   Describe two advantages of using 'biological' detergents compared with non-biological detergents.